How to Be Your Own Butcher

OTHER BOOKS BY THE LOBELS

MEAT. The Lobel Brothers.

ALL ABOUT MEAT. Leon and Stanley Lobel.

THE LOBEL BROTHERS' MEAT COOKBOOK. Leon and Stanley Lobel with Jon Messmann.

How to Be Your Own Butcher

Stanley, Leon,
and Evan Lobel

ILLUSTRATIONS
by Lauren Jarrett

A
Perigee
Book

Perigee Books
are published by
The Putnam Publishing Group
200 Madison Avenue
New York, New York 10016

Published simultaneously in Canada by General Publishing
Co. Limited, Toronto.

Library of Congress Cataloging in Publication Data

Lobel, Leon.
How to be your own butcher.

Includes index.
1. Meat. 2. Meat cutting. 3. Meat—Boning.
I. Lobel, Stanley. II. Lobel, Evan. III. Title.
TX373.L63 1983 664′.9029 82-19070
ISBN 0-399-50755-8

Designed by Stanley S. Drate
Edited by Jon Messmann
First Perigee printing, 1983
PRINTED IN THE UNITED STATES OF AMERICA
1 2 3 4 5 6 7 8 9

To the profession of my great-grandfather, my grandfather, my brothers, myself, and my son, to all those who have practiced this profession across the centuries, to the apprentices of today, and to all who want to and will learn the art of bringing the sustenance of life to the table, I dedicate this book. Anita, I love you.

—Leon Lobel

I dedicate this book with love:

To my wonderful wife, Evelyn, who makes me feel happy and alive. To my dearest sons, David and Mark, who make me feel blessed, and have given me reason to be proud. To my brother, Leon, with whom I have shared the past 25 years making our business a success. To my nephew, Evan, our newest generation in the business, who has come so far in so short a time.

—Stanley Lobel

I dedicate this book to Mom and Dad, who both helped, and let me be my own person. You're Number One in my book.

—Evan Lobel

Contents

Introduction

Meat and Your Table

The enjoyment of meat and the nutritional benefits it provides have long been a culinary foundation stone of American culture and American tastes. But today, modern Americans are bombarded by endless nutritional theories and programs for proper diet. There are theories on high-protein diets and on low-cholesterol intake, on vegetarian diets and yogurt diets, on the need for more grains and less frying. These theories come and go, often with every new piece of knowledge that arrives on the nutritional scene. The truth is that the study of what is best for you to eat is an ongoing search.

It is also true that man has been a meat eater from the earliest days of human existence. While nutritionists and others may debate about the value of grains and vegetables, many groups of people lived mainly on a single, high-protein food such as buffalo or caribou. The Arctic peoples have lived strenuous lives mainly on whale, seal, and fish. While such diets may lack variety, and perhaps certain nutritional elements, the people who subsisted on them were both strong and apparently healthy. The nutritional role of meat has been proven not only in the scientist's laboratory but, perhaps more importantly, in the laboratory of life.

Americans, however, always a nation of meat consumers, face a new set of situations in purchasing their main staple for today's table. This is an age when mass marketing exerts a growing force on you, the American consumer. Today, in this era of supermarkets and mass distribution, your meat comes to you, more and more frequently, in prepackaged form: cut, prepared, and portioned, sometimes directly from the packing plants, sometimes packaged by the regional supermarket operators. The old-fashioned butcher, that artist of the butcher block, grows more and more scarce. In some areas, he has all but disappeared.

And in this era of prepackaged meat, you, the consumer, are confined to taking those cuts and/or sections the prepackagers put in their market cases. Soon, if present trends continue, mass-market operators will completely decide what is to be offered and featured in a given week. Aided and abetted by their sales and specials, their purchases will govern your purchases. In some areas this is the situation today.

But there is a rebellion, or at least a quiet protest, taking place on the part of the American consumer. It is a rebellion based on economics as well as independence. The prepackagers have created a situation which makes more and more people fearful that their favorite foods will one day be available only in packaged form. Intelligent consumers want to be masters of their own tables. They want to have what they want when they want it. They want the freedom to have a particular cut of a particular meat at a time of their own choosing. They want to control their menus, not be at the mercy of a marketing specialist. That is the independence part of the rebellion.

Consumers are also attempting to do something about the high cost of meat, a cost which all the prepackaging techniques don't seem to lower. More and more, budget-minded consumers are purchasing their meat in bulk and learning how to make the best use of it. Buying your meat in bulk can very materially lower your food bill. For example, if you go to your local meat market and buy a shell steak and a piece of filet mignon cut from the porterhouse, the two pieces at the precut, prepackaged market price would cost you X number of dollars, allowing for the going market price. But if you bought these pieces in bulk and cut them at home yourself, *you would save approximately 25 to 35 percent* over the cost of those same pieces precut in your retail market. The same ratio, sometimes even a greater one, applies to other bulk purchases.

A center-cut beef chuck might cost you X dollars per pound when you cut it at home from your bulk purchase. The very same piece, marketed and prepackaged as boneless chuck and fillet of chuck at your retail market, would probably cost you double or more; for the same amount of money, you would have produced twice as much meat for your table by buying in bulk and doing your own cutting. Similar savings are there in almost every kind of meat when you learn to buy in bulk and to cut, bone, and trim it yourself.

We have been increasingly aware of how many people are frustrated by the growth of prepackaged meat distribution and alarmed by the dollars-and-cents cost of feeding their families. This book has grown out of that awareness. We want to help you, the American consumer, become more independent—both economically and culinarily—in this era of prepackaged meat.

There is no magic to this process. It is simply a matter of learning how to cut meat and practicing what you've learned. You will not become an expert butcher overnight, but then you would not expect to become skilled overnight at anything worth learning. However, when you have mastered the art—and it is an art as well as a skill—you will be able to cut what you want out of the meat you've bought in bulk and to experience the great pleasure of having the dishes you want on your table when you want them.

But the rewards are not only in surmounting the prepackaged world and its restrictions, not only in the substantial dollars-and-cents savings. There is another reward: the satisfaction of creating your own meals from almost the very beginning to the finished, steaming-hot dish. It was that way once. The great chefs of history all knew how to cut their own meats, and as a matter of course, they passed this knowledge on to their apprentices. Learning to cut meat for yourself and your family is a feeling not unlike that of having your own garden, of taking part in the basic growing process, of providing for your family and not being constrained by the whims and profit motives of shippers, market experts, artificial techniques, prepackagers, and all the other elements in the retail marketing pyramid.

To prepare your own meat will let you touch hands across time with other generations, to know that satisfaction which once belonged to your grandparents and everyone before them. Frozen and prepackaged foods have only one reward—speed. Many Americans are deciding that is not nearly enough.

In the presentation of this guide to being your own butcher at home, we show those techniques which the Lobels have developed, refined, and found to be best for home use. In our shop in New York, we like to believe that our clients receive, for their money, meat cut and trimmed, prepared and offered, with the skill and artistry that enhance the finished product. Craftsmanship, like everything else, exists on many levels. We hope that some of this attention to craftsmanship will be passed on to you, our readers, through these pages. But we have kept in mind that there are certain procedures that are generally beyond the province of anyone except a highly trained butcher and not generally suitable for home use. The techniques outlined here are those we feel anyone can learn and, with practice, skillfully perform in a home setting.

Though each section—beef, veal, chicken, etc.—may be studied separately and is complete in itself, we have begun the book with those areas we feel are easiest for the newcomer to the butcher's block. We then proceed to the next step up in the order of skill required. In this way, your own skills and proficiency will keep pace with the difficulty and complexity of the subject matter, cuts, and techniques explained.

This modest volume, then, is to help you, the American consumer, save today, prepare for tomorrow, and recapture some of yesterday.

—THE LOBELS

1

Buyer Tips

Buying meat in bulk from a distributor or wholesaler is a very different process from going to your local supermarket. In the various chapters we speak of the elements of conformation, the signs of quality, and other factors to be borne in mind. It is equally important that you be informed properly on the subject of grading.

First, there are two kinds of grading, one for quality and the other called *yield grading*. In the retail market, shoppers are concerned almost wholly with quality grading. As a purchaser of bulk meat, you must be concerned with both. But even in the matter of quality grading, recent changes in government regulations mean that most people have new information to learn. There are still 8 quality gradings for beef: USDA Prime, USDA Choice, USDA Good, USDA Standard, USDA Commercial, USDA Utility, USDA Cutter, and USDA Canner. In the retail supermarket you would probably find only the first or top three grades: Prime, Choice, and Good. In the bulk-purchasing markets or wholesale distributors, you will probably find most of the 8 grades, and you should thoroughly familiarize yourself with the meaning of these labels.

Prime is the top-graded meat and usually the most costly. It is also the most flavorful. Produced in smaller quantities than the other grades, Prime ages best because of its thick fat cover. Choice is the next grade, and the most popular one in overall consumption and retail sales. *Good* is inferior to the other two grades but by no means unpalatable. Meat officially graded Good is often the meat which, in the retail market, bears the store's brand name. Standard falls slightly below Good and if it appears in retail markets is often sold without either a grade or a house-brand name.

Commercial, Utility, Cutter, and Canner, which almost never appear in retail markets, are grades bought by companies engaged in processing meat for sausage, cold cuts, luncheon meats, and various other meat products. These grades of meat are usually ground, precooked, or otherwise treated to make a meat product. Yet at a bulk-purchase distributor you may find beef with all of these quality gradings, and it is important that you pay attention to them.

To further complicate the business of buying the meat you want, recent regulations now allow beef which was formerly graded Choice to be graded

Prime, based on the decision that much grain-fed Choice beef fits the Prime qualifications. Furthermore, there are now variations within the gradings: Top Prime, Middle Prime, and Low Prime; Top Choice, Middle Choice, and Low Choice; Top Good, Middle Good, and Low Good. Top Good is sometimes called Fancy Good.

Various factors are taken into consideration by the graders when making a determination of quality. The amount of marbling (intermingling of fat with lean) is one important factor. Though these requirements have been altered slightly in the newest quality gradings they are still used a great deal. There are 9 marbling levels in the official Grader's Manual: Abundant, Moderately Abundant, Slightly Abundant, Moderate, Modest, Small, Slight, Traces, and Practically Devoid. Because the government regulators have devised such a carefully evaluated grading system, the importance of grading to the consumer cannot be overemphasized. Yet even so, you should ask questions about the beef your marketer handles: where it comes from and the prevalent feed used. In general, grain-fed Western beef is still considered the top-quality beef on the market.

Of equal importance to you when you are buying bulk meat for your home butcher block is that previously mentioned designation of grading: yield grading. Entirely apart from quality grading, yield grading is a procedure that determines roughly how much edible meat in proportion to bone and fat a given carcass will supply. Obviously, this is of major importance to the bulk buyer. Yield grades are numbered from 1 to 5 and are so stamped. Yield number 1 will provide the most edible meat from a given carcass, number 2 the next amount, and so on, with number 5 yielding the smallest amount of edible meat compared to fat and bone. Yield grading was adopted to encourage beef producers to use new feeding, raising, breeding, and handling techniques in the production of cattle, bringing meatier cows to the marketplace with more tender and flavorful qualities.

However, you, the buyer, must decide on how you can best use yield grading. Price, personal tastes, family size, and other factors will eventually allow you to use yield grading in the manner best suited to your needs. For example, it is obvious that a piece of Choice marked Yield Grade 1 will give you more edible meat than a piece of Choice marked Yield Grade 3. However, a piece of Choice Yield Grade 2 may be a better buy because of cost per pound for a given family than a piece of Prime Yield Grade 1, even though the Prime will yield more edible meat. Conversely, a buyer may prefer a Prime Yield Grade 3 over a Choice Yield Grade 1, choosing the Prime carcass for superior flavor despite the smaller amount of edible meat involved. Personal preferences and budgeting considerations will and should determine how you use the yield-grading system. It is an important tool when you buy meat in bulk.

There is still some confusion about the term Inspection. This has nothing to do with the grading systems. Inspection to assure cleanliness is required by federal law for every meat packinghouse, processing or selling outlet. State and municipal laws are usually also involved in such inspection, and although these vary from locality to locality, all must either meet or improve upon the nationwide federal standards. Simply put, *quality grading* is for your palate, *yield grading* for your pocketbook, and *inspection* for your health.

In the matter of grading, it should be noted that nutritional differences in protein, mineral, and vitamin contents vary very little among the top three quality gradings. Good will have more moisture and less fat than Choice or Prime, because the lean meat of Good will have less marbling. Lean muscle contains more water than fat muscle. Because of this, the leaner Good-graded meat may well shrink more in cooking, and this may offset the economy offered by the lower price. Lastly, always take into account the basic quality grading of your meat when cooking. Different grades of beef require adjustments in cooking methods and times. This point will be amplified in Chapter 8.

There is one more important element to bear in mind when purchasing your meat in bulk. You must be prepared to transport it home cleanly and efficiently. Most suppliers will furnish the wrapping for your bulk purchases. However, you might check on this with your individual market. Bulk purchases can be wrapped in plastic, though plastic is slippery and may be difficult to handle. Heavy-duty Kraft paper or burlap may also be used.

After you have gotten your bulk purchase home, it is essential that it be refrigerated quickly. You should be prepared to cut and package your meat at once or to divide it into sections you can refrigerate until you have time to cut your meat into suitable pieces either for immediate use or for freezing. Do not allow the meat to lie about unrefrigerated. Spoilage can begin quickly under many conditions. Therefore, plan ahead when making bulk purchases. Be ready to go to work on cutting and boning, or be sure you have enough refrigeration space waiting to hold your purchases until you begin this work.

A sculptor needs the proper tools to carve, mold, and chisel. A painter needs an array of brushes with which to paint. A mechanic needs the right tools to fix a car properly. Indeed, the finished result is often governed by the tools used. It is no different in the cutting and preparation of meat. One could rightly say that the taste is partly in the tools because the precision with which you cut, trim, and prepare the various cuts of meat will affect the final taste on your dinner table.

You do not have to buy a voluminous array of knives, shears, cleavers, and saws. Some people do amass a supply of all shapes and sizes of cutting tools, but the truth is that you need only a few basic tools to do the tasks described and pictured in this book. These few tools will be used 90 percent of the time. They are: 1. a good butcher's saw; 2. a cleaver; 3. a 12-inch butcher or chef's knife; 4. an 8-inch knife.

These are your basics. There are also 14-inch knives available, 6-inch knives, paring knives, and various other implements that you may find convenient and useful. But to do the various tasks outlined here, you need no more than the few tools mentioned above. However, there are some facts regarding your knives which you should know before making your purchases. On today's retail market, there are three basic blade materials offered to the purchaser. Each has some advantages and some drawbacks. The basic blade materials are:

Carbon steel: Carbon steel can be honed to the sharpest cutting edge and it holds the best edge. However, carbon steel will discolor and rust if not dried thoroughly after use.

Stainless steel: Stainless steel will not discolor; it resists corrosion, and is stronger than carbon steel. It does not take on as fine an edge as carbon steel and requires more frequent sharpening. This, however, is not really a drawback because you will be sharpening your blades frequently anyway.

High-carbon stainless steel: This is a combination blade being made more and more frequently by manufacturers. It will not rust and takes a sharper edge than carbon steel. However, sharpening this blade requires that you have a special, extra-hard sharpening steel. We shall discuss the matter of sharpening knives later in this chapter.

There is one more material used for knife blades that is made by a few manufacturers. This is *super-stainless alloy steel.* Blades made of this steel are highly touted as never needing sharpening. Understandably, the advertisers neglect to inform purchasers that the blade is so hard it is virtually impossible to sharpen and there are no sharpening steels or stones hard enough to sharpen it. Super-stainless alloy knives, once their edge is gone, can be thrown away.

The "Rockwell Hardness Test" is a method used to rate the hardness of metals. A conical, diamond-pointed weight is dropped from a prescribed distance, and hardness is determined by the depth of the resulting dent. Most butcher's knives are rated in a range of 54 to 56 on the Rockwell Scale. Sharpening steels—because obviously the sharpening material must be harder than the blade—are rated between 64 and 68 on the Rockwell Scale.

As you shop for knives in the general market, you will find certain brand names most often sold in cutlery departments. You should know that these knives often vary in the way they are manufactured—even those blades produced by a single manufacturer. Some blades are stamped steel, others are of one-piece forging, and some are of two-piece forging. A one-piece fully forged blade is usually the finest in quality and strength. So, in addition to inquiring about the kind of steel in a knife you are considering, you should also ask about the way the knife has been made.

It would be impossible to list here all the brand names you may encounter when buying your knives. Some blades are made by companies that deal almost solely in trade sales to butchers and wholesalers. Other companies sell to both trade markets and the general market. Just to serve as a loose guide, we list here a few of the brand names to look for.

Sabatier: This is probably the most famous of the brands in the general market. Sabatier makes both a carbon-steel blade and a high-carbon stainless steel blade, and you should specify which you prefer. Furthermore, Sabatier knives are produced in some 14 different factories in France, have differing stamped imprints, and differ as to quality. The "Lion" Sabatier stamp is probably the most consistently reliable of the Sabatier brand.

Wusthof: This is possibly the finest of chef's knives made today and usually the most expensive. It is a one-piece, fully forged high-carbon stainless steel blade, made in Germany.

Henckel: This knife, also made in Germany, ranks close to the Wusthof in quality, but it is of two-piece forging and, therefore, of slightly less tensile strength. It is also a high-carbon stainless steel blade.

Chicago Cutlery: This is an American-made knife of carbon stainless steel. Until recently, Chicago Cutlery made only professional knives to be sold in trade circles, but they have now entered the general market. Less expensive than the first three names above, the blade is stamped, not forged.

Case: This company, also American, probably produces the most economical of all the blades listed here. Case makes a variety of stamped stainless-steel blades sold in retail outlets.

Copco: This manufacturer makes a forged blade of high-carbon stainless steel, with good tensile strength and with a great deal of attention paid to details of interest to the design-oriented buyer.

It is wise to ask for, and insist on, full information from a knowledgeable seller when buying knives, because all companies change their manufacturing systems and policies from time to time. You should also know something about handles. Today, handles come in wood, plastic, laminated combinations, and rubberized-plastic materials. Wood, particularly Brazilian rosewood, gives the best grip. Many plastic handles are too slippery when in use. Polypropylene and other rubberized materials wear well and need less care than wood. You should look at the shape of the handle and hold it in your hand. Some shapes are more comfortable than others. Buy the one that feels best in your own hand.

Part of the blade itself extends into the handle. This is called the *tang*. As a general rule, the finer the knife, the fuller the tang, which means that, in better knives, the tang goes fully through the handle. You can usually tell what kind of tang a knife has by inspecting the handle. A full tang is secured by three rivets in the handle, and the edge of the blade is visible along the bottom and top of the entire handle; a three-quarter tang can also have three rivets visible, but the blade is not visible along the lower edge of the handle; a half tang is usually marked by only two rivets in the handle and the blade extends only partway into the handle.

Along with the proper knives, it is very important to acquire the right tools for sharpening them. You simply cannot do the job of cutting meat, of being your own excellent butcher at home, with dull knives. Also, dull knives make it necessary to use increased pressure, and this in turn increases the chances for injury. So sharply honed knives are a must. With that in mind, we find that we cannot recommend many forms of sharpeners available today: metal disks, electric sharpeners, wall holders, and abrasive wheels. These devices may suffice for sharpening a vegetable-paring knife but not for the knives you will use as you become your own butcher. We recommend either a honing stone or a sharpening steel.

Honing stones come in various materials. They also come in grades of coarse, medium, and fine. Carborundum stone is probably the most common, but there are also stones composed of silicon carbide, sandstone, aluminum oxide, and other abrasive materials. Coarse stones are, naturally, for use when a knife is in poor condition. We feel you should never need a coarse stone, because your knives should never be allowed to get into poor condition. Medium honing stones are the most widely used, and you may well need no other than this, but the fine grade can give a particularly sharp edge.

There are various ways to use a honing stone. All require pulling the blade over the stone to sharpen the edge. Use the method most comfortable for you, but there are a few points that apply to all.

First, the sharpening stone should be securely held down, either in its box or on a damp towel, to prevent it from moving. Next, hold the knife blade at a constant-angle, approximately 20° to the stone. Then, always sharpen in the same direction. Finally, use the same number of strokes on both sides of the blade.

The most frequently used method is to hold the knife by its handle and draw it gently but firmly along the flat top of the stone, the blade angled to create the abrasion needed to sharpen the edge. Some butchers use their other hand to help guide the blade along the stone, using a light pressure with thumb and two fingers atop the side of the blade. Others simply draw the blade along the stone using one hand only. Drawing the blade over the stone should be done in one smooth continuous action, not with little short strokes.

When the blade is sharpened, carefully wipe off the edge with a damp towel to remove any residue.

The sharpening steel is not so much a substitute for the honing stone as it is an additional tool to give a really fine edge. It is used after the honing stone has done its preliminary work. As with the stone, there are different techniques used in handling the sharpening steel. All involve stroking both sides of the blade against the steel. One method recommends holding the sharpening steel horizontally, an arm's length from the body, and passing the knife blade along the steel almost at a right angle. Another consists of placing the tip of the steel down on your cutting board, making sure it does not slip—if necessary placing a damp towel on the board to insure this—and drawing the blade down the length of the steel, first one side, then the other.

We have found that holding the steel in one hand at approximately a 45° angle, and the knife in the other, drawing the blade first down one side of the steel and then down the other, seems to be the most natural action. You will find your own most comfortable angle and method for using the sharpening steel, but however you use it, keep these points in mind: One, feather the blade against the steel, a light stroke, not a heavy grinding action. This is most important. Listen as you stroke. You should hear a light ringing tone. (If you hear a heavy, grinding noise you are stroking with too much pressure.) Two, make the strokes evenly on both sides of the blade. Three, do not overstroke—five to ten strokes per side should be more than enough.

Remember, the sharpening steel, not the honing stone, is the tool to use to keep restoring the fine edge to your blade while you work. You will be using the sharpening steel a lot more frequently than the honing stone. Therefore it is important that you get the right steel, one in which the handle fits comfortably in your grip, and with a good protective guard.

Sharpening steels are probably the most important tool you will need for your knives. Again, the shaft of the sharpening tool must be harder than the caliber of the blade. Therefore, it is important that the sharpening steel (and/or the honing stone) be purchased from a reliable dealer. It is even more important to tell the dealer what kind of steel blade you are using. Sharpening steels come not only in several degrees of hardness but also in a variety of grains. Some have two grains combined in one steel on different sides. In choosing your sharpening steel, also pay attention to the handle and the guard. As with the knives, handles come in various shapes, and, as

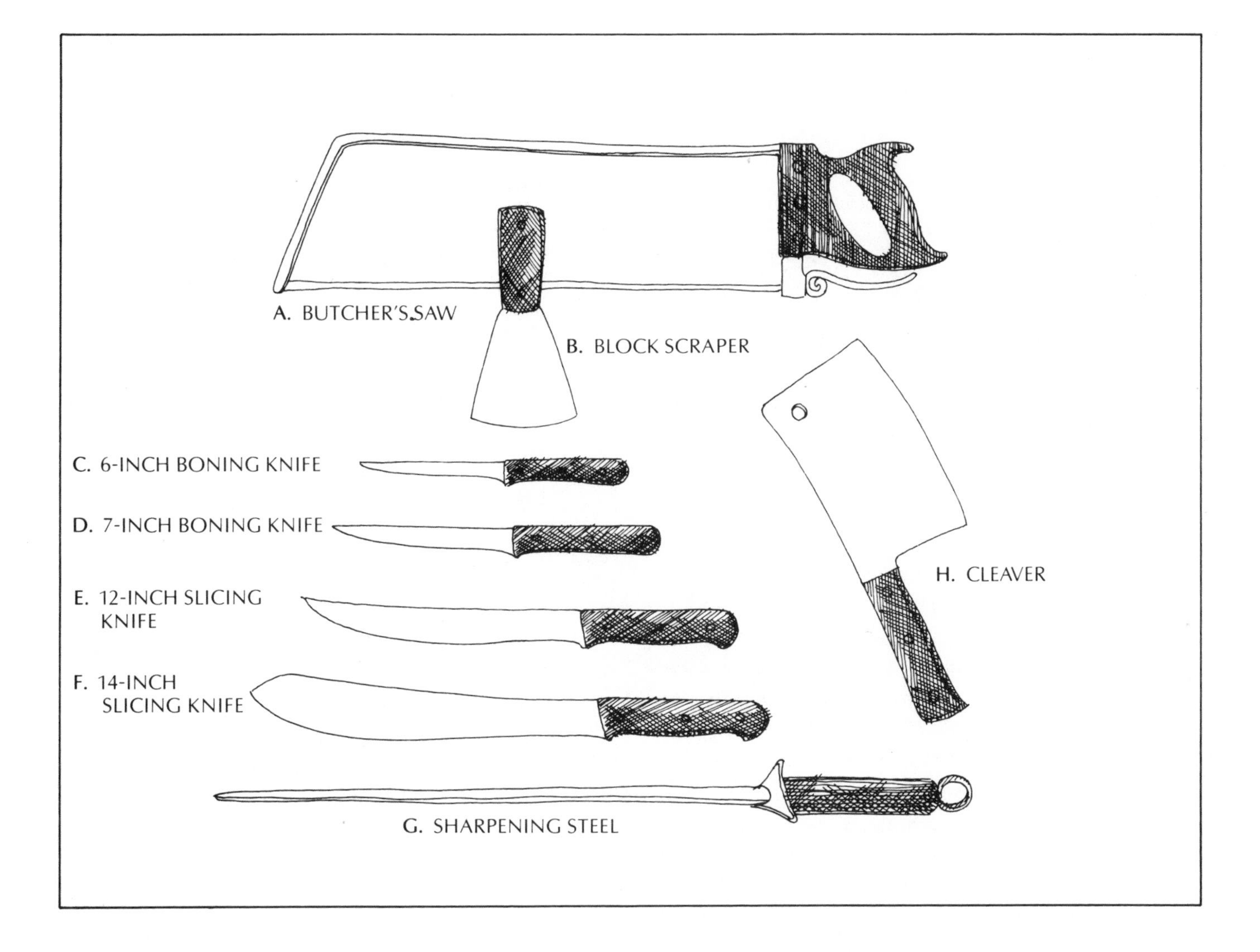
A. BUTCHER'S SAW
B. BLOCK SCRAPER
C. 6-INCH BONING KNIFE
D. 7-INCH BONING KNIFE
E. 12-INCH SLICING KNIFE
F. 14-INCH SLICING KNIFE
G. SHARPENING STEEL
H. CLEAVER

we stressed earlier, you should choose one that feels comfortable to you. The handle guard is most important because you will be sharpening your knives against this steel. Moreover, you will be doing this many, many times, often briefly in the midst of your cutting to acquire a slightly sharper edge. The guard should be large enough to protect *your* hand against an occasional slip.

There are certain rules and work habits that you should observe. Among the rules are:

—Never put knives in a dishwasher.
—Do not soak knives in dishwater. It is not good for the blade and can be a dangerous trap for the unwary.
—Never use knives to open bottles or pry up lids or for other tasks for which they are not intended. Use them strictly for cutting.
—Always wash and dry knives when finished with them.
—Store knives in a safe place at all times.

Work habits should include these important practices:

—Never run your fingers along the edge of a knife to test its sharpness.
—Never try to catch a knife if you drop it. Let it fall and get yourself out of the way.
—Wear footwear strong enough to protect you if a knife should fall on your foot.
—Don't hand a knife to another person. Set it down and let the other person pick it up.
—Never use the cutting edge of the knife to scrape away trimmings or crumbs.

As to tables and cutting surfaces, wood is still unquestionably the best. A standard butcher's block or counter can be bought from those stores that sell to the trade. Metal tables are definitely not recommended. They will injure tools and meat will tend to slip on them. The resilience of wood designed as a cutting surface permits chopping and cutting of meat without splintering or excessive wear of the surface. Ceramic tables are also hard on tools and on your arms. Wooden surfaces and/or tables should be scraped and cleaned daily and sandpapered frequently. After wood, the next-best cutting surfaces are those made of hard rubber and plastic, which can be easily cleaned. However, these are not usually made wide enough and long enough for heavy-duty cutting.

In storing the saw and cleaver, sturdy wall hooks on which they can hang safely will usually do quite well. Knives should be stored more carefully. Keep them out of the reach of children, of course. A cork over the point of the blade is important if the knives are to be hung. The wooden knife block, where the blades are securely nested in their slots inside the block, is the safest. There are also slanted blocks that are made to fit under a counter or sink top.

A word about saws. You should purchase the handsaw known as a butcher's handsaw. These are best ordered from supply houses for the trade and/or for restaurant equipment. Under no circumstances should a saw made for cutting metal be used. The saw will not require sharpening as your knifes do. When the teeth finally do wear down, it is best to replace them or to replace the entire saw. For cleaning, all that is needed is a damp cloth rubbed carefully over the sawteeth to remove any particles sticking there. Dry after rubbing clean.

Cleavers should also be ordered from trade supply houses; purchase the one that feels most comfortable in the hand. As with knives, a good nonslip handle is necessary.

For sewing pockets in meat, any sturdy needle with a large eye may be used. We suggest the use of a larding needle, if possible. Any sturdy twine may be used, but we suggest #16 butcher's twine as the best for sewing meat.

In the matter of clothing, aprons are recommended for wear when cutting meat. Any good working apron, even the colorful plastic type, is adequate. The typical white butcher's apron is less stiff than the plastic apron, therefore more comfortable to wear for long periods. In general, gloves are not recommended. They decrease the sensitivity in the fingers and hands, a very important element in good meat cutting. Shoes should be of the nonskid variety—with sturdy tops if possible—capable of allowing you a surefooted grip on the floor.

It is, as you see, important to use good tools for your work. Whether you purchase a commercial name brand or a brand sold primarily to professionals through trade outlets, the care of your tools is of equal importance. And, we reiterate, a sharp blade is a must in order to do a good job. That is the first commandment of every professional butcher and, indeed, of every craftsperson who uses tools that require sharpening. Make it yours.

3

Lamb

Lamb is one of the most versatile of meats and lower in calories than beef or pork. Because lambs can be raised under poorer conditions than beef cattle, lamb has been for countless centuries the staple meat for most of the Middle East, North Africa, and the Mediterranean basin.

Lamb is particularly responsive to subtle seasonings and spices. Most of the lamb you will purchase fresh in the United States will be from 6 to 9 months old (except baby lamb, which is from 4 to 6 weeks old). Lamb, to be called by that name here, must be under 12 months old. The grade should be marked prime or choice high-grade and should have aged from 1 to 2 weeks.

Lamb "dresses out" at approximately 50 percent, which means that an 80-pound lamb will weigh in as a 40-pound carcass. Trimmed leg roasts will weigh from 6 to 8 pounds each; shoulder roasts from 5 to 7 pounds; breast and neck approximately 6 to 8 pounds. There will be approximately 6 to 8 pounds of loin and rib for roasting or for cutting into about 20 to 25 good, thick chops.

Remember, lamb should never be overcooked; it dries out easily. It is best when faintly pink on the inside. Do not roast till it is brown-gray in color with the meat falling from the bone. Roast only until the juices run pink, approximately 20 minutes per pounds at 325°F.; and simmer slowly in liquid at 25 minutes per pound.

MAJOR SECTIONS OF THE LAMB

Cut individual pieces from the following major sections of the lamb:

THE SHOULDER

If a recipe calls for the following, cut them from the lamb shoulder:

Shoulder Roast (also called Square Cut Shoulder; Shoulder Block)
Boneless Shoulder Roast (also called Rolled Shoulder Roast; Boneless Shoulder Netted)
Shoulder Blade Chops (also called Blade Cut Chops, Shoulder Lamb Chops)
Arm Chops (also called Round Bone Chops; Arm Cut Chops)
Neck Slices (also called Lamb Neck for Stew; Lamb Stew Bone In)
Lamb Stew Meat (also called Boneless Lamb Stew; Stewing Lamb)

THE BREAST

If a recipe calls for the following, cut them from the breast of lamb:

Breast of Lamb
Lamb Riblets (also called Breast Riblets)

Lamb Spareribs
Stuffed Breast of Lamb

THE RIB

If a recipe calls for the following, cut them from the rib:

Lamb Rib Roast (also called Lamb Rib Rack; Hotel Rack; Rack Roast; Rack of Lamb)
Lamb Rib Chops (also called Lamb Rack Chops)
Lamb Rib Crown Roast (also called Lamb Crown Roast)
Frenched Rib Chops (also called French Lamb Chops; Rib Kabobs)

THE LOIN

If a recipe calls for the following, cut them from the loin:

Loin Roast (also called Saddle of Loin; Saddle Roast)
Laced Saddle of Lamb
Kidney Chops
Loin Lamb Chops
Loin Double Chops (also called English Chops)
Boneless Double Loin Chop
Boneless Loin Rolled Roast (also called Rolled Saddle; Double Loin Roast)

THE LEG

If a recipe calls for the following, cut them from the leg:

Leg of Lamb (also called Haunch of Lamb; Leg Roast; Leg, Sirloin-On; Full-trimmed Leg Roast)
Shank of Lamb (also called Leg, Shank Half, Foreshank)
Lamb Butt (also called Lamb, Sirloin Half)
Frenched Lamb Leg
Boneless Roast Leg
Sirloin Chops
Butterflied Leg of Lamb
Kabobs
Lamb Scaloppine

WHAT TO LOOK FOR WHEN BUYING LAMB

When buying lamb at the market, look at the entire lamb, whether you plan to buy the whole piece or only a portion. It is most economical, of course, to buy the entire lamb—and a whole lamb isn't too large to handle.

A well-bred, tasty lamb will have certain physical characteristics. Many ancient Persian recipes begin with the words "Look for a fat-tailed lamb." Those ancient cooks knew what they were talking about. You may not be able to find a tail, fat or otherwise, on your lamb, but a fat-tailed lamb has a fat, chunky rear and hindquarters that are short-legged. Compare the lambs you see. Many, if not most, will be long and narrow in body with long hind legs. Look for a lamb with a rounded midsection—a full-bellied, compact animal. The edges of the breast, on both sides, should be firm. There should be a uniform, cream-white coating of fat over the lamb. The outside color of the lamb should not be dark or grayish.

Look inside the dressed carcass. The color should be a bright pink or red, not dark, dull red. The fat around the kidneys should be thick and creamy-white, not yellow. Compare the lambs you see. Note the differences in the thickness of the bones. The thinner the bones, the younger and more desirable the lamb.

You may not always find the fat-tailed, chunky-reared, short-legged lamb you want, but you now know what to look for and to buy when you can find it. Outer conformation, color, and texture indicate the taste of the meat when it reaches the table.

FIRST STEPS: SHOULDER WITH FORESHANK ATTACHED

You have brought your lamb home (L–1). The first step is to cut off the section called the shoulder (or chuck) with foreshank attached. This is the section with the stubby base of the neck attached at the top and the forelegs at the bottom; it is the front part of the lamb.

Starting at the base of the neck, find the first, shortest bone and count down 4 or 5 bones. Make an incision, using a 5- or 6-inch knife (though you will use the saw for the major cut), between the 2 bones where you have halted your count. Saw straight *across* the carcass, from one side to the other (L–2).

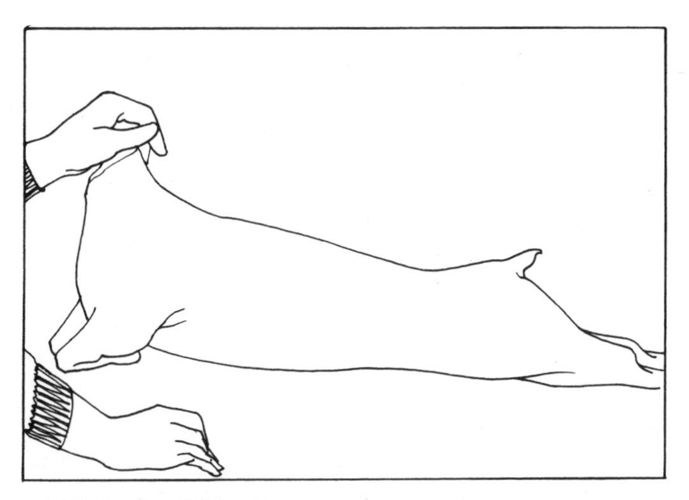

L-1. THE ENTIRE LAMB

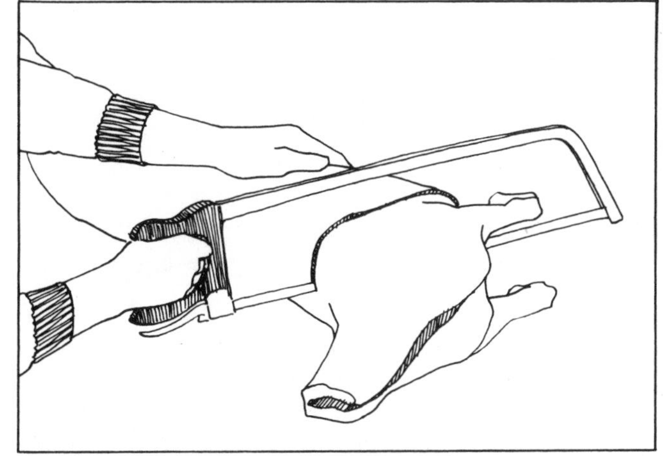

L-2. CUTTING OFF FORESHANK SECTION

Never press too hard on the saw—let the blade glide. When you finish, you will have the entire section with the foreshank attached.

This piece is then cut down the middle (L–3). It is best done by a band saw, but it can be done with a regular handsaw. Cut straight down the backbone, along the chine bone (the chine bone is the top of the backbone).

You now have 2 halves, right and left. Each half is now cut into 3 pieces (L–4): the foreshank, the neck, and the main part or shoulder proper.

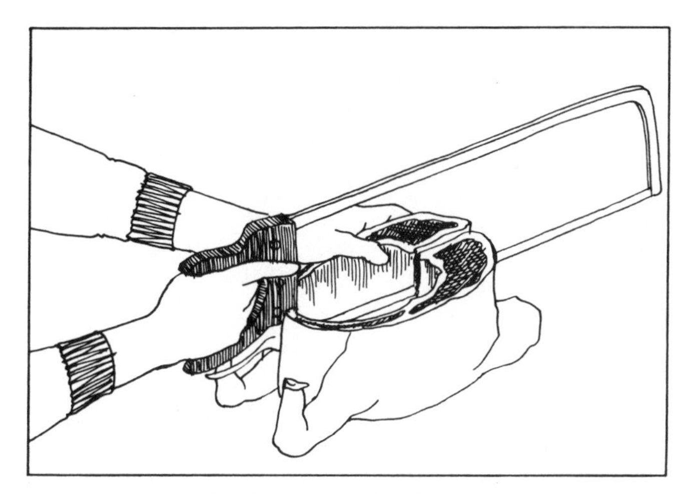

L-3. CUTTING FORESHANK IN HALF

ARM CHOPS AND BLADE CHOPS

Out of this shoulder section you can cut 2 round-bone (arm) chops and 4 blade chops. Simply cut down through the bones all the way and you have your chops. Trim fat from the chops around edges and they are ready to cook (L–5).

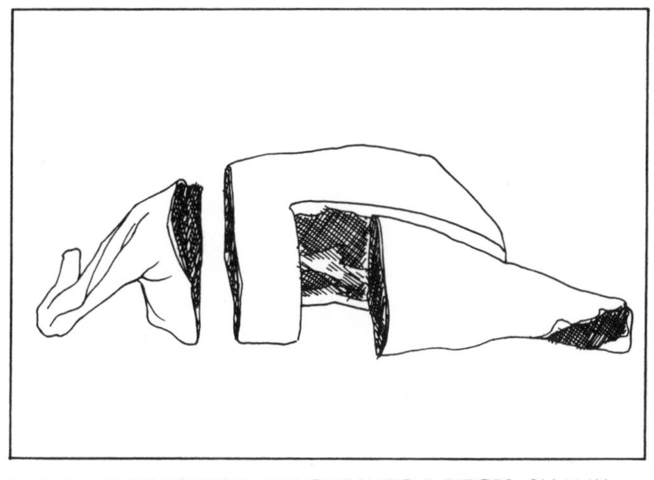

L-4. HALF OF FORESHANK CUT INTO 3 PIECES: SHANK, SHOULDER AND NECK

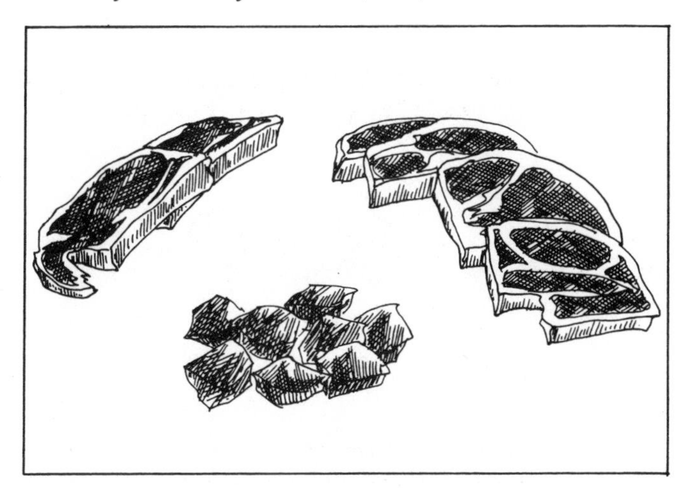

L-5. 2 ARM (ROUND-BONE) CHOPS, 4 BLADE CHOPS, AND LAMB STEW PIECES

STEW MEAT

For stew, trim excess fat from the foreshank and the neck, cut off the meat underneath, and cut the meat into medium-sized cubes. You can make a more elegant stew by cutting the meat of the chops into cubes. But using the chops as chops and the neck and foreshank meat for stew gives you the most for your money from this section of the lamb.

SHOULDER ROAST

For the shoulder roast, first saw off the foreshank and the neck piece as shown in L–6. When finished, you will have separated a squarish piece of meat from the foreshank and neck pieces. This is your shoulder roast. (We must note here that this is not our preference for a roast. We feel it is too fatty and too bony to slice. Some people do favor it, however, and some recipes call for it.) Trim the outside fat from the square shoulder roast piece. Cut deeply into the pocket of fat next to where the neck was joined and remove this. Turn the piece over and crack the bones (L–7) so they separate but do not come apart entirely. Turn over again and crack the bones from the top side—4 cracks (L–8). Tie the roast as shown in L–8. It is now ready for cooking.

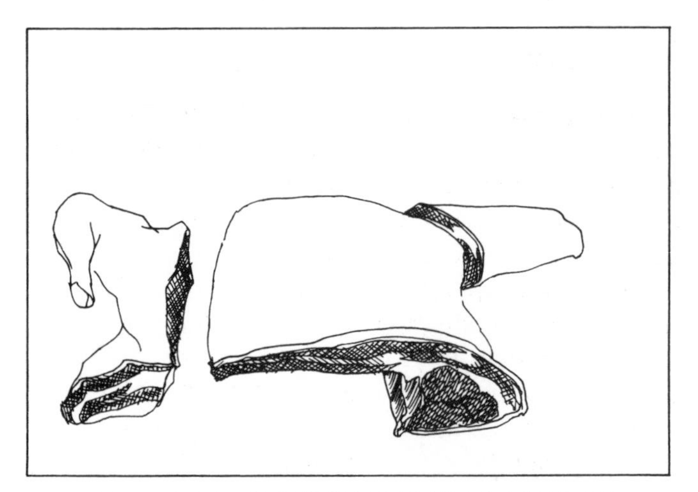

L-6. TAKING SQUARE-CUT SHOULDER ROAST FROM 2ND FORESHANK HALF

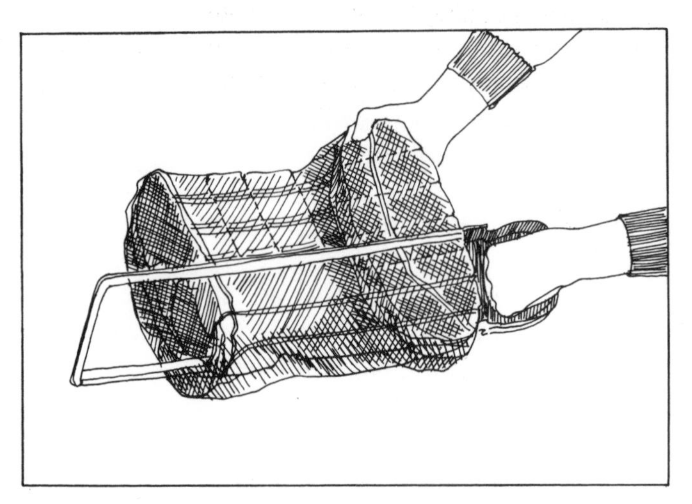

L-7. CRACKING SHOULDER-ROAST BONES

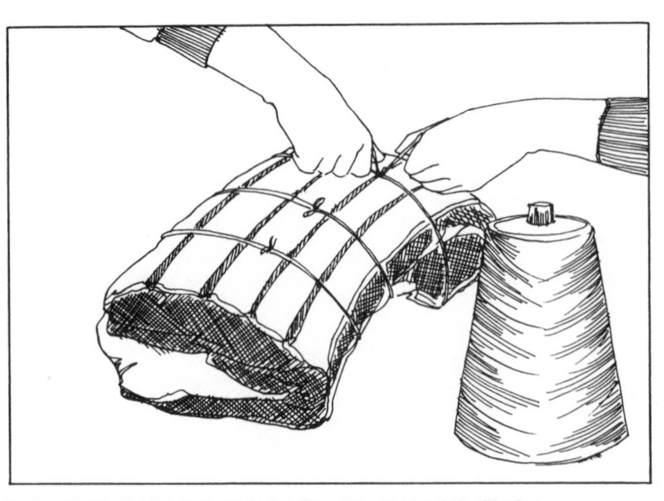

L-8. TIED SHOULDER ROAST, READY FOR OVEN

BONELESS SHOULDER ROAST

To bone, take out the underside rib bones with the 8-inch knife (L–9). Entirely remove the rib bones and the neck-bone pieces still attached. Then remove the blade bone (L–10) by making a slit under the outer fat and cutting around the two sides of the blade bone. At the deep end of the blade bone you'll find a smooth round bone, which also must be entirely removed. However, this round bone can be removed separately by cutting around it and pulling it free. The entire boning operation is a difficult one, but it can be done if you take it slowly, step by step.

With rib bones, neck pieces, blade, and round bone removed, the shoulder roast is boned and ready to be rolled. First, remove the pocket of fat as you did when cutting the shoulder roast with bone in; then roll from the two shorter ends of the piece. Tie strings crossways over the roast when finished rolling, and it is ready for the oven (L–11).

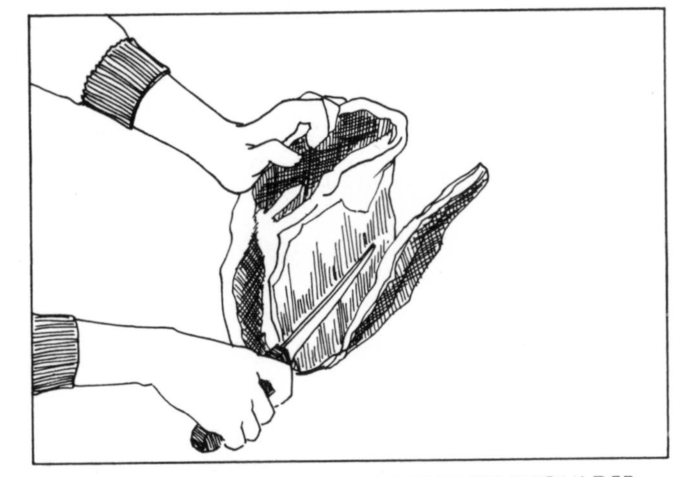

L-9. REMOVING RIB BONES FOR BONELESS SHOULDER ROAST

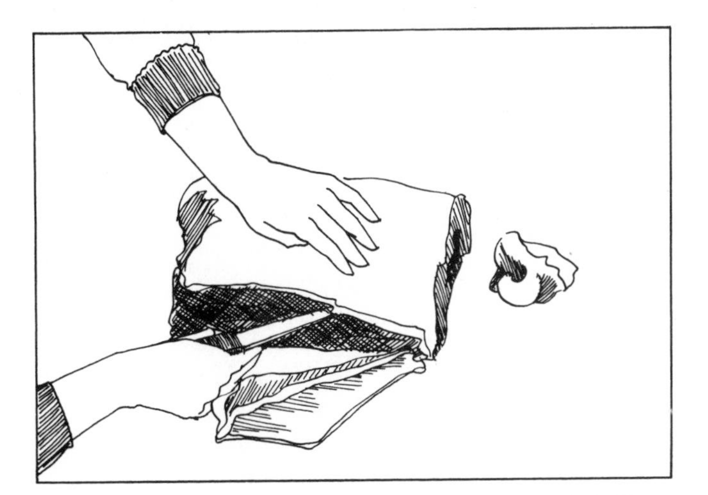

L-10. WORKING KNIFE AROUND BLADE AND ROUND BONE

L-11. FINISHED BONELESS ROAST

THE BREAST

Once the foreshank section is removed, what is left is a piece composed of the breast, ribs, and hindquarters. To remove the breast, first determine its size. Do this by measuring along the chine bone (the top of the backbone) approximately 5 inches. With a saw, cut this 5-inch width parallel to the backbone (L–12), then cut at a right angle to remove one side of the breast. Do the same with the other breast. You will have 2 long, flat breast pieces. These will include the lower part of the rib bones.

Turn the breast pieces over and prepare each the same way. First, remove the very thin membrane over the meaty part of the bones. Then, turning the breast over again, remove the *fell* (outer skin), a very thin, wax-paperlike tissue of skin. Do this by holding the knife flat and gliding it along just under the fell, cutting very shallowly as you lift the fell. When you've removed the entire fell, cut away the visible fat just beneath the outer skin (L–13).

Crack the bones in between the rib sections with the cleaver. This is breast of lamb. It may also be used as spareribs (L–14).

RIBLETS

For riblets, simply cut through the sections of bones altogether.

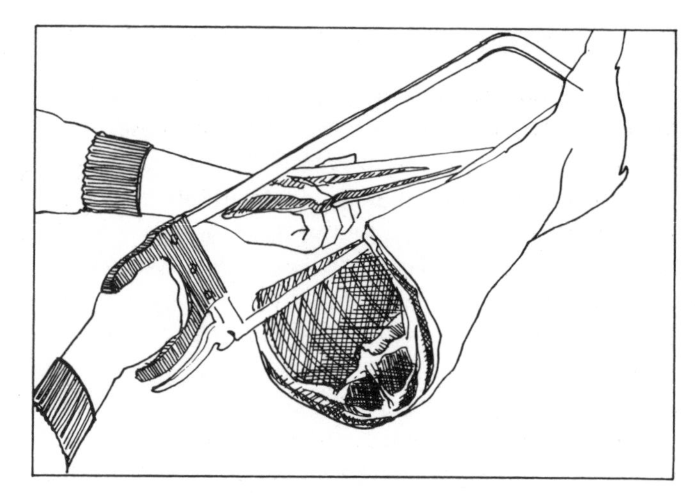

L-12. CUTTING BREAST OFF RIB SECTION

L-13. TRIMMING BREAST SECTIONS

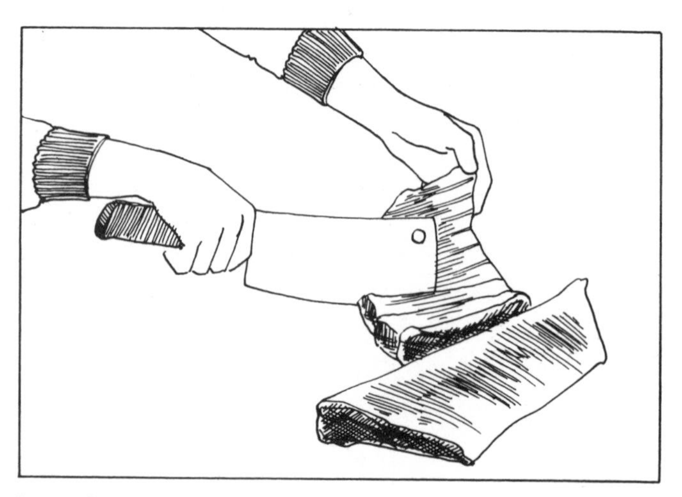

L-14. CRACKING BONES BETWEEN RIB SECTIONS

STUFFED BREAST OF LAMB

For stuffed breast, place the point of the knife at the end of the thicker part and slip it into the breast. Guide the incision down toward the bones so you *do not* cut through the surface of the meat. You must use a 6-inch knife for this, one that can cut flat and make a long incision. Move the blade lightly back and forth in the breast until you have made a slitted pocket, keeping a firm grip with the other hand at the other end of the breast (L–15). When you have finished cutting the pocket, and have made it deep enough, turn the breast piece over and, very lightly, crack the cartilege between the bones. Stuff the pocket with whatever stuffing you choose.

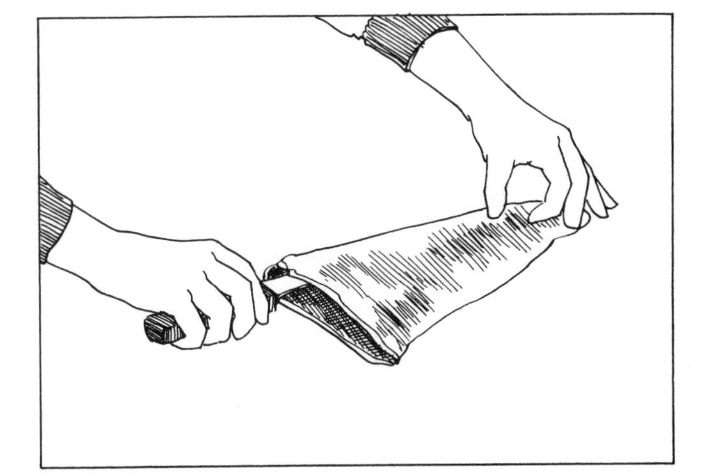

L-15. MAKING POCKET FOR STUFFED BREAST

LAMB KIDNEYS

After you have cut out the breast sections, the lamb kidneys will be visible inside the lamb between the loin and the rib sections (see L–12). They will be surrounded by thick fat. Remove the fat and take out the kidneys to use as you wish.

THE RIB

The rib section must be cut from the loin section. To do this, cut all but the very last rib bone, which will be left on the loin. With the lamb on its back, feel the bones with your fingers until you find the next to the last rib bone. Using your knife, cut between these two bones down to the backbone. *Caution:* You can't simply cut straight down. The bones curve on a bias, and you must watch your cutting and follow the curvature of the bones.

Cut to the backbone only (L–16), then turn the lamb on its side, and using a saw, saw through the backbone—but only through the backbone itself. Then, again using the knife, cut through the other side of the piece, once again in the space between the last two rib bones.

The finished section will be a half-moon-shaped rib section.

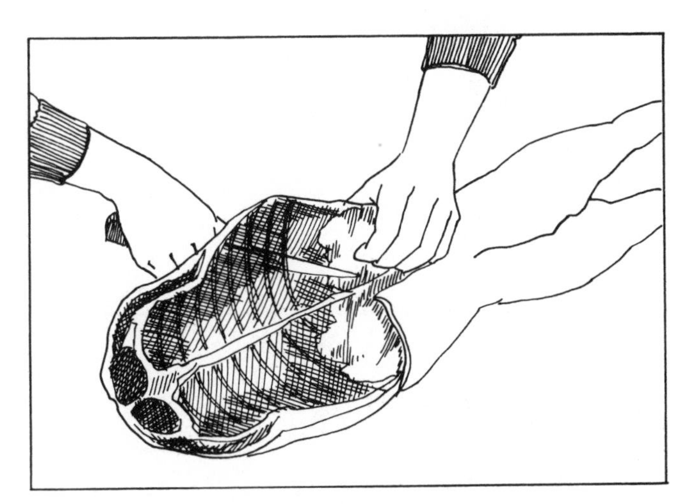

L-16. REMOVING RIB SECTION FROM LOIN

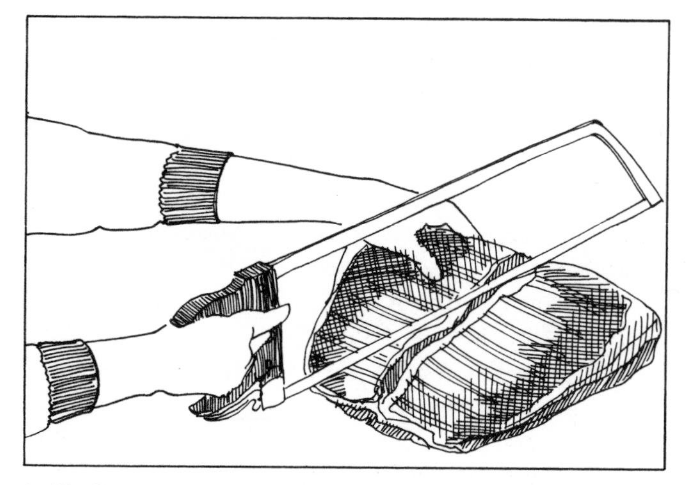

L-17. CUTTING RIB SECTION IN HALF

LAMB RIB ROAST OR RACK

Using a saw, cut this half-moon rib section in half along the backbone (L–17). You will then have 2 racks—but they are not yet oven ready. First you must remove the backbone, using your saw. Then, cut off the long tips of the rib bones, cutting against the bones. Remove the chine bone (L–18), along with the piece of cartilege attached, using your knife. Then, slice off some of the outside fat (L–19). You will have a rack of lamb. Stand it on the bottom end so that it curves upward to the thin bone top and, with a knife or saw, cut between the bones but do not cut all the way through (L–20).

To make the rack more elegant, you can "french" it by making a cut 1 inch from the end of the bones (L–21), at a right angle to the bone, around the bone itself. Then cut down that inch between the bones, working the knife to strip away all the meat on that one inch until you have only an inch of bare bone sticking out at the end (L–22). At the end of the roasting period, slip a paper ruffle or cuff on each bone end (L–23).

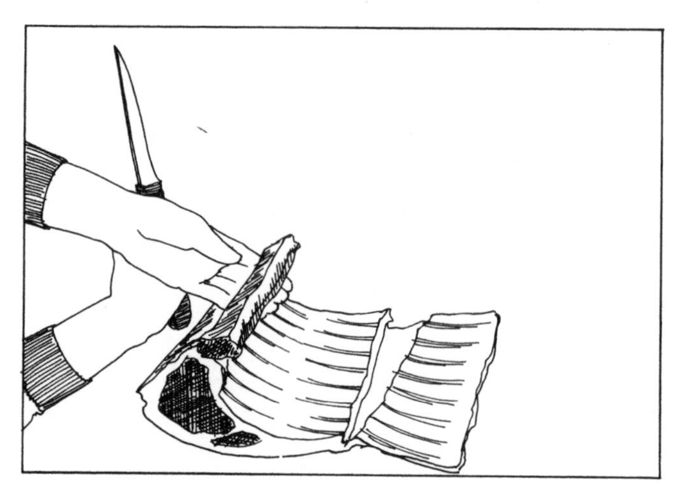

L-18. REMOVING CHINE BONE AFTER SAWING

LAMB CHOPS

Prepare the rack as described above but do not cut off the top length of the bones. Instead, cut all the way through between the rib bones for regular rib lamb chops.

FRENCHED LAMB CHOPS

Cut chops as above and then french the bone at each end as described under Lamb Rib Roast (L–23).

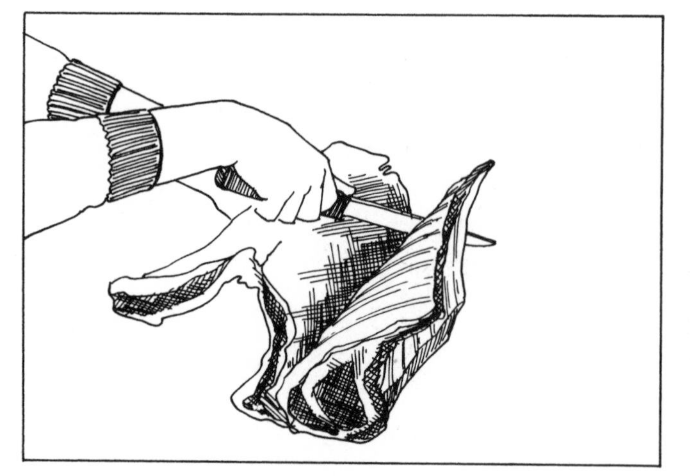

L-19. REMOVING TOP OF RIB AND TRIMMING OUTSIDE FAT

L-20. RACK OF LAMB: CRACKING BETWEEN BONES

L-21. STARTING TO FRENCH THE RACK OF LAMB

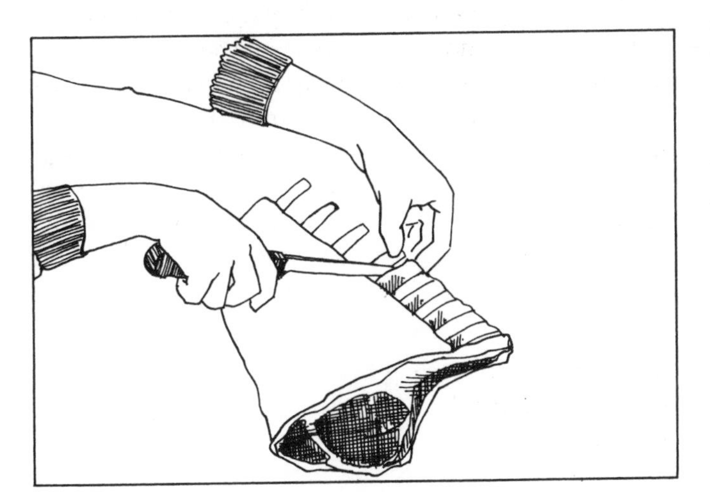

L-22. FRENCHING

CROWN ROAST

Using the saw, cut down the center of the backbone to cut the entire rib section into 2 racks, as you did for the rib roast. However, this time *do not* remove the backbone. Trim the fat from the bones on the *inside* of the racks. Remove all meat on the top of the rib bones and then trim off the excess fat left after removing the meat. French the top of each bone as described earlier (L–23). Measure and be sure that the frenched tails on both racks are the same length. *Do not* cut down between the bones as you did with the rib roast.

Now there are two racks with the bone ends frenched. With chine bone up, on each rack, make saw cuts across the chine bone, placing each cut between the rib bones. *This is critical.* Saw into the chine bone just far enough to keep it on the meat, or until you just reach the "eye" of the meat (L–24). Stop sawing the moment you cut this far through the chine bone. Now, changing back to the knife, continue cutting, using the tip of the blade, into the backbone and ¼ inch into the meat (L–25). But again, not too deeply—just enough so that the cuts between each rib bone and through the chine and backbone will allow you to turn the rack inside out. *Do not cut all the way through.*

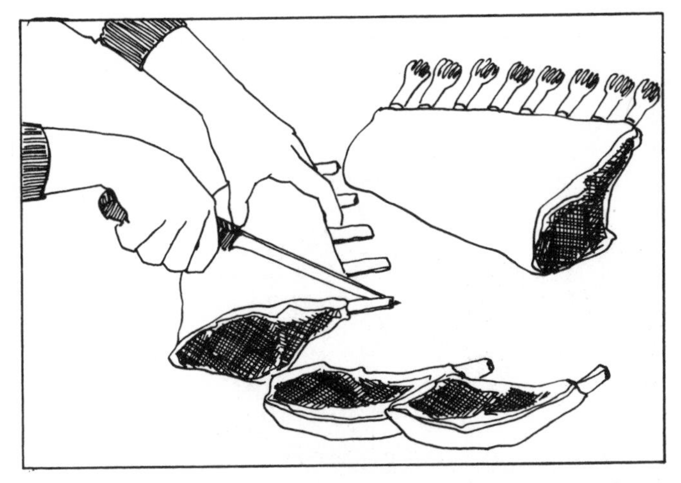

L-23. SINGLE FRENCHED CHOPS AND FRENCHED RACK

L-24. SAWING CAREFULLY INTO CHINE BONE FOR CROWN ROAST

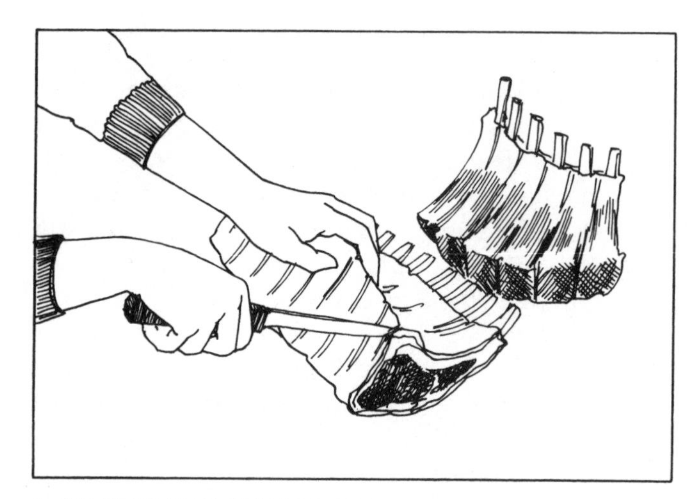

L-25. FINISHING CUT FOR CROWN ROAST

L-26. USING NEEDLE TO TIE CROWN ROAST SECTIONS

When this has been done with both racks, turn them inside out—here are the two parts of the crown. Place them together and lay the crown on its side with the backbone facing you. The 2 bones which are not end to end will be tied together. Do this by taking a heavy needle and pushing it through the meat just under the backbone. Put string into the eye of the needle and withdraw the needle. The string should be through both bones to connect them. Tie the string, securing the bottom of the bones together. Now tie the 2 bones at the top of the crown together in the same way (L–26). Now form a circle by bringing the remaining open sides together. Repeat the tying operation, first under the backbone, and then at the top, as with the first joining (L–27). Cut away the loose ends of the strings, and you have your crown roast. Place a circle of aluminum foil at the bottom of the center of the crown (to prevent the stuffing from dropping out). Now fill the crown with the stuffing of your choice. Leave the top of the stuffing rounded. When the crown has finished roasting, add paper ruffles or cuffs to the ends of the bones (L–28). Olives, cherries, or kumquats may be used instead of paper cuffs to add an especially festive touch.

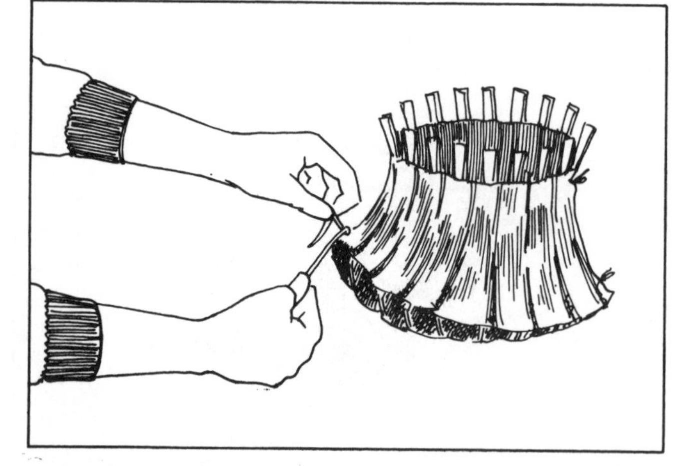

L-27. FINISHING THE TYING ON OTHER END OF CROWN ROAST

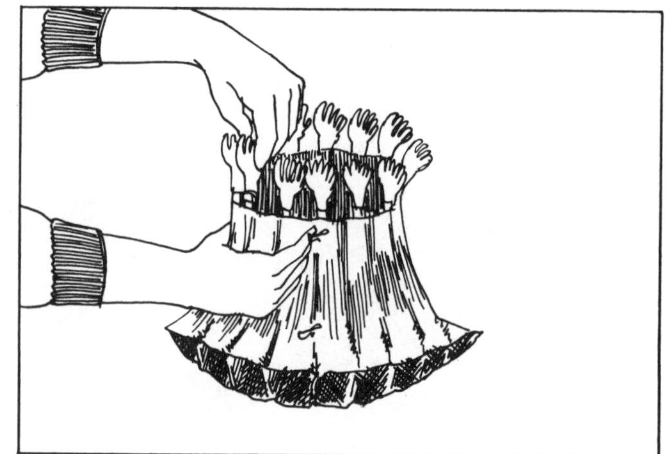

L-28. ADDING FRILLS

THE LOIN

First separate the loin from the leg. Run your thumb down one side of the lamb until you feel a large bone. This is the hipbone. Make a cut with the 14-inch knife just in *front* of the hipbone (L–29). Remember, the hipbone will stay attached to the leg. Then, using a saw, saw straight down through the lamb. When this is done, you will have a half-moon double section. This is the full or double loin.

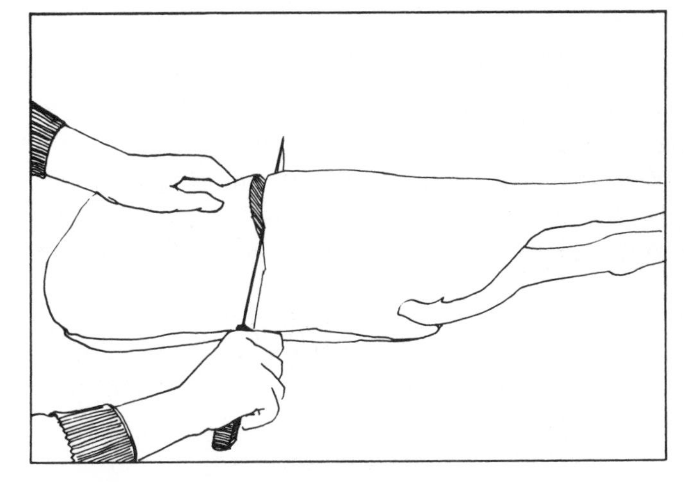

L-29. FEELING FOR HIPBONE AND CUTTING LOIN FROM LEG

SADDLE OF LOIN

This is the traditional loin roast. First remove the fell from the outside of the lamb (L–30) as for the rib roast. Then trim the inner parts of the saddle. Remove the last rib bone which was left on when you cut the rib section away. Trim away any excess fat on the inner parts of the saddle (L–31). Tuck the bottom loose parts under the main portion to make a neat double saddle (L–32) so that the backbone runs through the center of the roast. You now have a full-trimmed saddle, bone in, ready for the oven. No tying is needed.

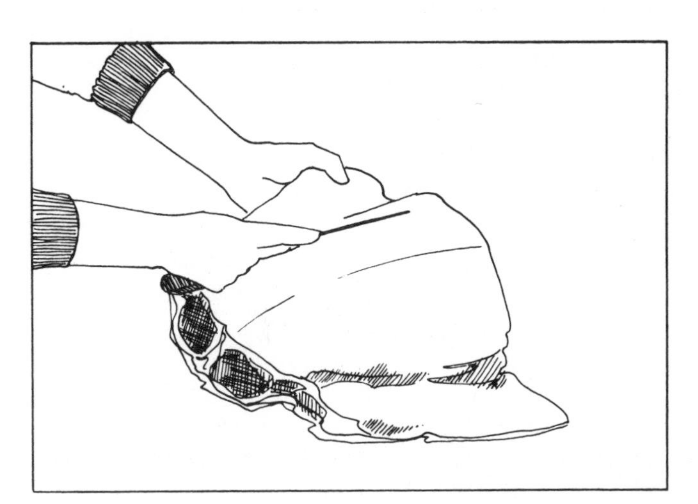

L-30. REMOVING THE FELL

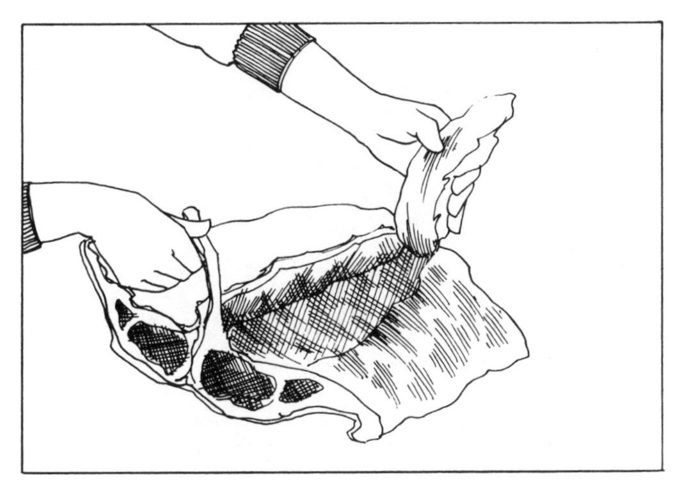

L-31. REMOVING EXCESS FAT ON INNER SADDLE

L-32. THE SADDLE OF LAMB

LACED SADDLE OF LAMB

Along the full saddle with the fat side on top facing you, make 2 cuts, one on each side of the backbone (L–33). Cut down until you reach the chine bone. Work the knife until you free the meat from each side of the chine bone. Now there are 2 long, funnel-like cuts, one on each side of the backbone. Work the knife further until the cuts are flexible and opened enough (L–34) to fill each long cut (which should run almost the length of the saddle) with stuffing or pâté. After filling the cuts, tie string around the saddle in at least 2 places to keep the cuts closed after stuffing. The laced saddle of lamb is now ready for the oven.

BONELESS SADDLE

Spread the saddle out with the outside down. Trim the inside (i.e., remove final rib bone and excess fat), then gingerly cut with your knife to the end of the backbone (L–35). Take care not to remove the fillet, which nestles right against the backbone.

Run the knife gently under the backbone and continue to cut, following the contour of the T-bone; cut under but not completely through (L–36), and free it from the meat. Cut along the bone until it is entirely free and can be lifted out easily. Roll the ends up and tie. This is the boned saddle of lamb.

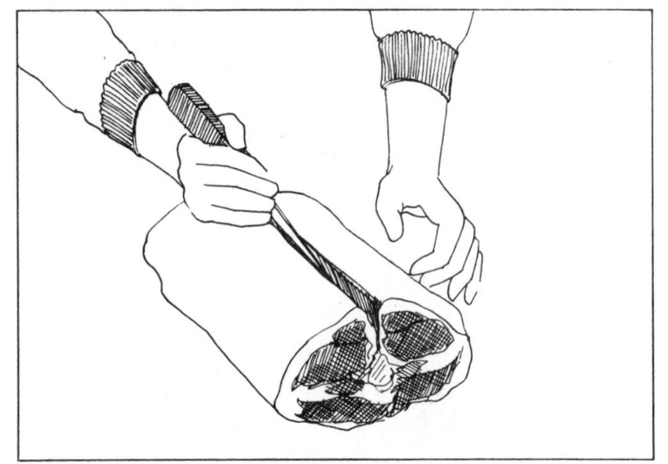

L–33. CUTTING ALONG BOTH SIDES OF BACKBONE

L–34. OPENING FOR LACING

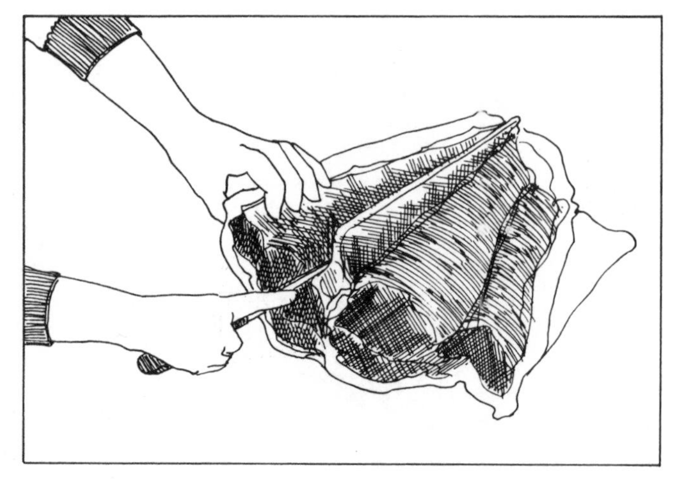

L–35. BONELESS SADDLE: BEGINNING TO BONE

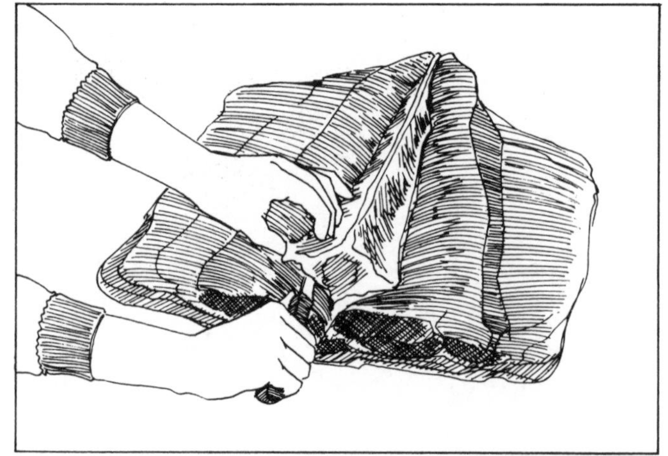

L–36. WORKING KNIFE ALONG BACKBONE

BONELESS STUFFED SADDLE

Follow the procedure for preparing the boneless saddle. When you've finished, lay the boned meat out flat (L–37). Place stuffing in the center where the backbone was removed, between the 2 fillets and on each side of each fillet. Carefully fold up one flank, then fold the other flank over it. Tie or skewer together (L–38).

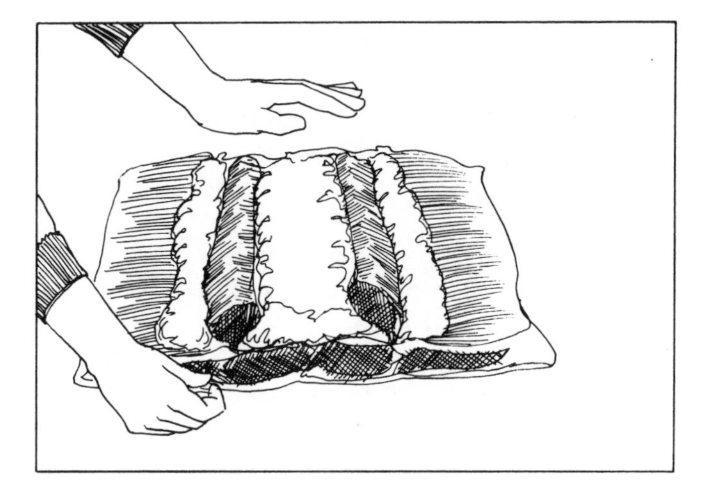

L-37. SADDLE LAID OUT FLAT AND STUFFED

BONELESS LOIN LAMB CHOPS

Simply cut the boned saddle into 1-inch-thick slices (L–39) or thicker if you wish.

KIDNEY LOIN LAMB CHOPS

Place 2 to 4 kidneys *inside* the boneless loin roll. When you cut the chops there will be a piece of kidney in each chop.

REGULAR BONE-IN LAMB CHOPS

Simply cut and saw through the bone (L–40).

L-38. TYING BONELESS SADDLE

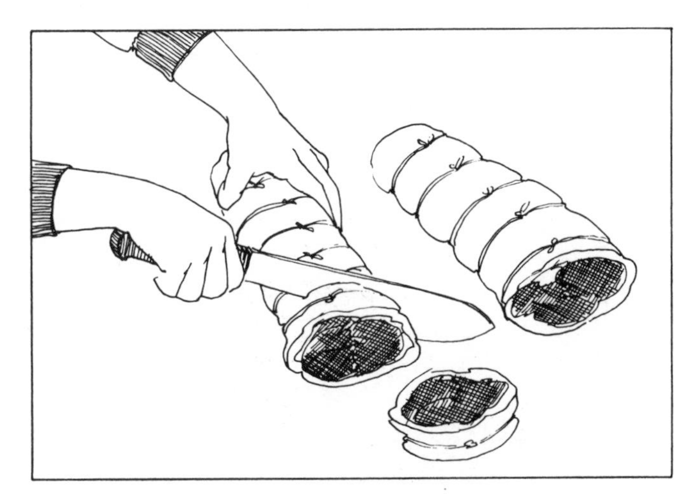

L-39. CUTTING BONELESS LOIN LAMB CHOPS

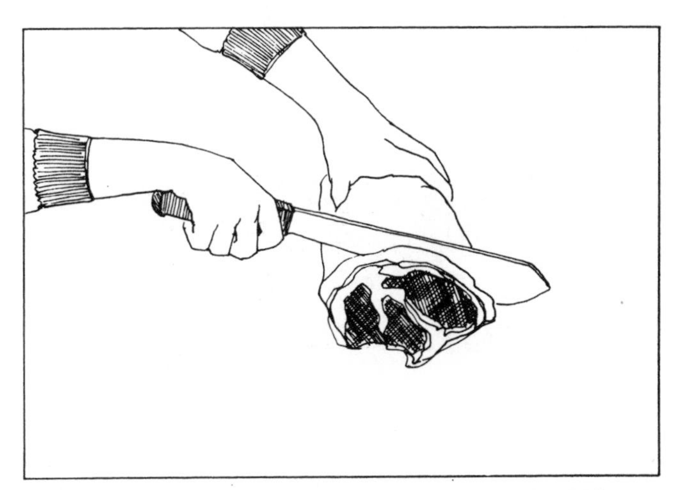

L-40. CUTTING 1-INCH REGULAR LOIN LAMB CHOPS BEFORE BONING

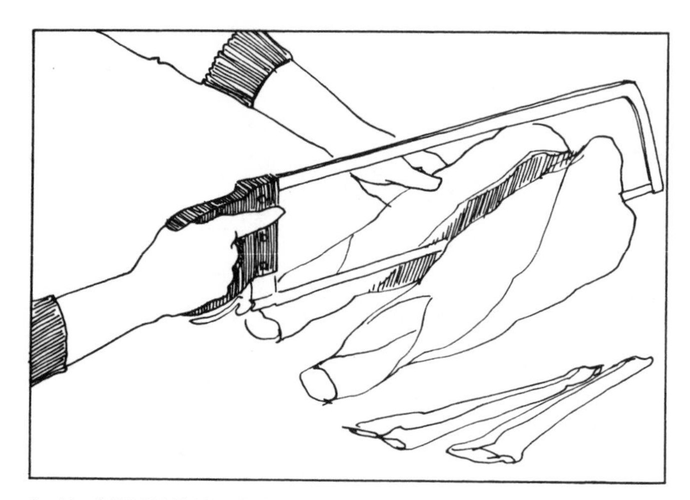

L-41. SAWING HINDQUARTER IN HALF

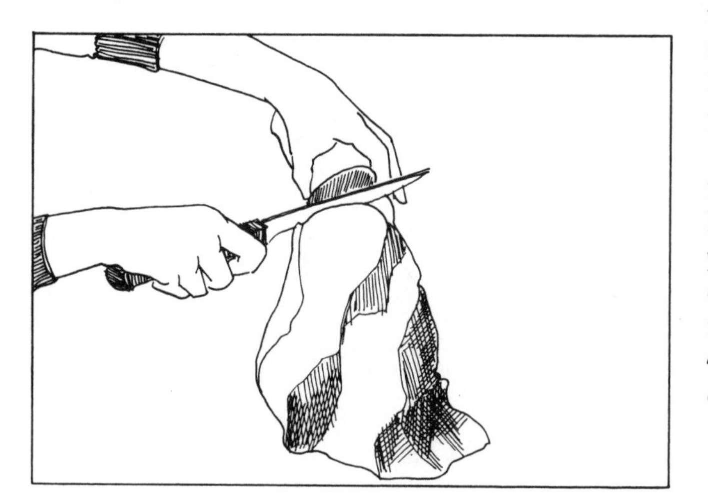

L-42. CRACKING AT KNEE JOINT

THE LEG

First saw the hindquarters in half (L–41). You will have two legs of lamb. Remove the bottom of the leg bones (shanks), just above the hock where the meat begins. Remove the last vertebra (tail), which you should save for stock.

LEG OF LAMB

With the inner side of the leg facing up, remove all the hanging fatty and gristly pieces. Then remove the thin membrane and continue trimming off excess fat until you are down to the meat. Turn the leg over. Work the knife down toward the end of the bone, taking off the tough top skin. Pare all the skin from the bottom of the leg. Turn leg over so that top is now face up. Start removing the fell from the top by loosening the outer corners. The fell can now be pulled off by hand.

On this part of the leg, keep the good, clean, fresh fat that remains after the fell is removed. Move the leg at the lower joint to feel the shin joint. When you've located it, crack it at the joint (L–42). For easier carving, also crack the upper end of the leg by making 2 or 3 cuts, 1 inch apart, with your saw (L–43). Score fat if you like (L–44), though this is not essential.

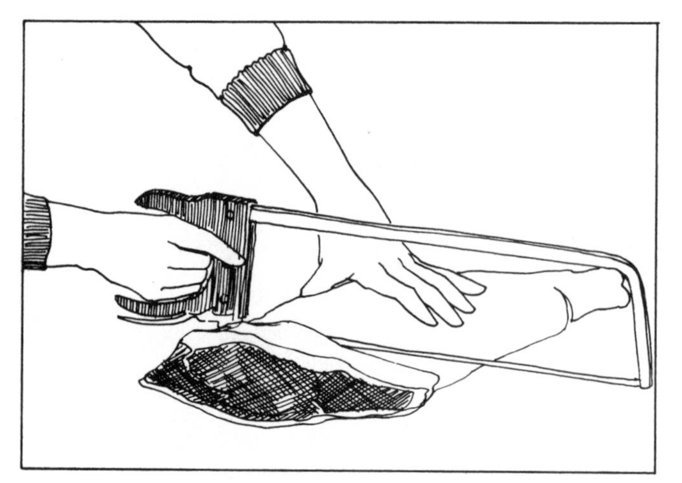

L-43. CRACKING 2 CHOPS

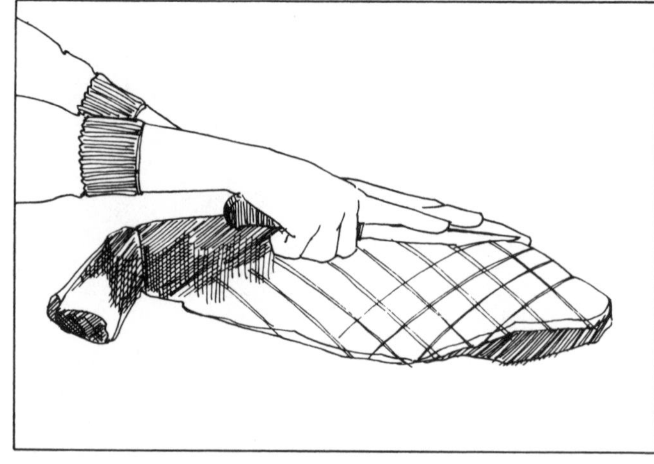

L-44. SCORING FAT

FRENCHED LEG OF LAMB

This is prepared exactly as above, except that it is *not* cracked at the shin joint. Instead, the lower end of the bone is bared, as with frenched chops. The bare end of the bone is then used as a handle when carving.

LAMB SHANK

Cut the trimmed leg across in half for lamb shank and sirloin section. It is from this hind-shank section that you get the center-cut steak (L–45) and the meat for kabobs (which will be described later).

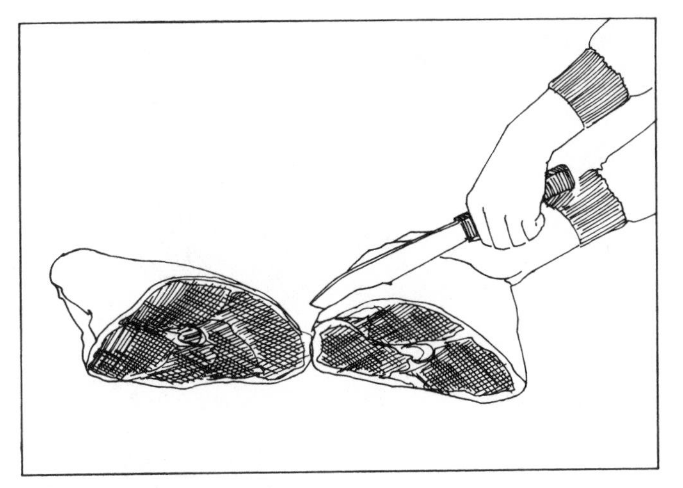

L-45. LEG CUT IN HALF

BONELESS LEG OF LAMB

Prepare the leg in the same manner as for the traditional leg of lamb. When this has been done, place the top side down so the bottom of the leg is facing you. You will be looking at the hipbone (L–46). Near the hipbone is a very smooth, round knob of bone. This is the leg bone. A cartilege connects the leg bone to the hipbone. Sever this, separating the two bones.

With the knife, cut around the hipbone. Then, cut alongside, lengthwise along the backbone, until the entire hipbone is freed and can be lifted out (L–47).

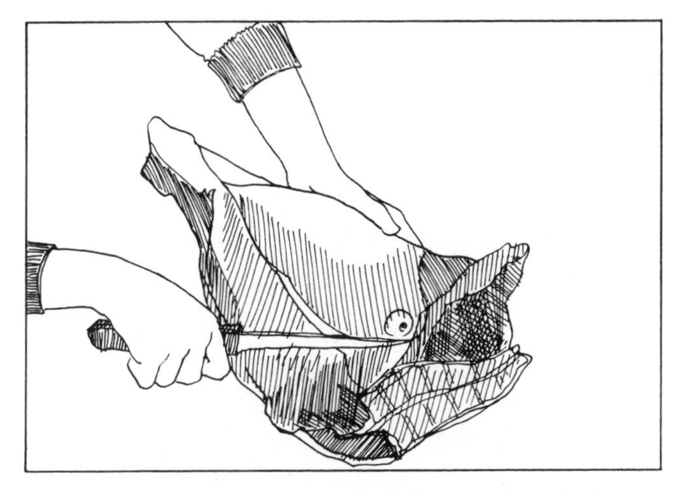

L-46. BONELESS LEG: CUTTING AROUND THE HIPBONE

The next step is to remove the round-knobbed leg bone. To do this, move the joint at the other end of the leg to find the shinbone and sever the skin at this joint. After you do this you'll be able to see the line of the bone between the shin end and the round-knobbed end. Cut along this line with the knife, down to the bone. Then work the blade around the bone to free it from the meat (L–48).

Glide the knife carefully. At the round-knobbed part take extra care, for the bone has other knobby ends which will become apparent as you cut around it. When this operation is completed and the bone removed, you will find that there is still a small section of bone left at the other (shin) end to take out. When you've removed this (L–49), the boned lamb is ready to spread out.

When the lamb is spread out, there will be a triangular, white fat gland (L–50) just up from the shank part. Remove this and roll up and tie the boneless leg of lamb (L–51).

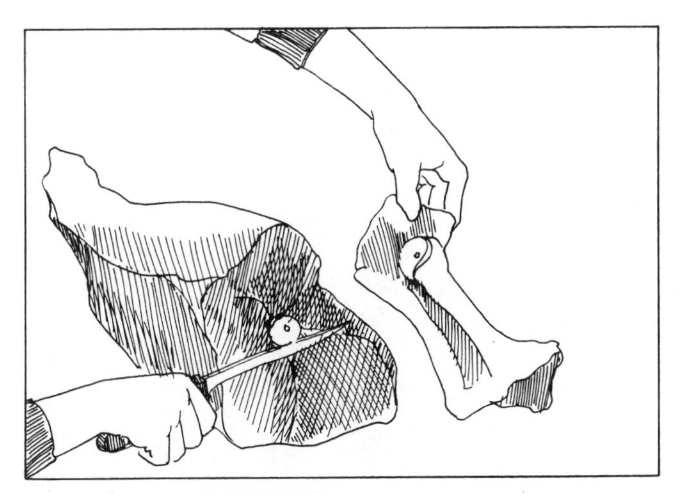

L-47. REMOVING HIPBONE

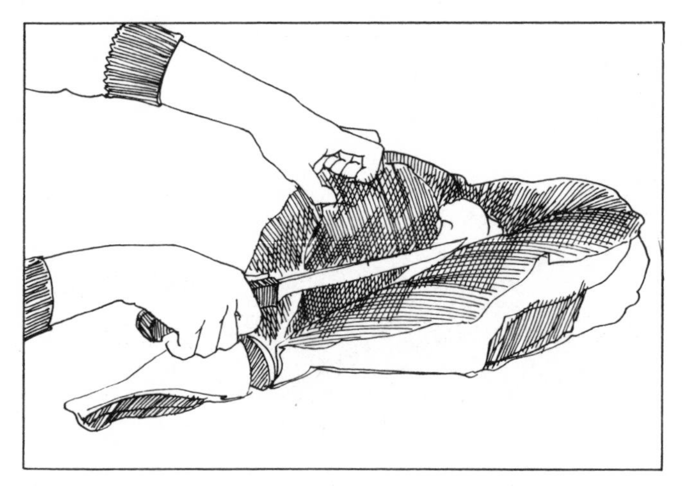

L-48. CUTTING TO REMOVE LEG BONE

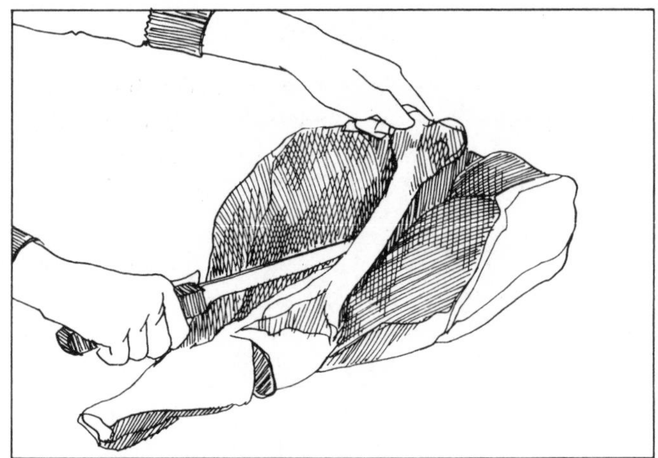

L-50. TAKING OUT FAT GLAND

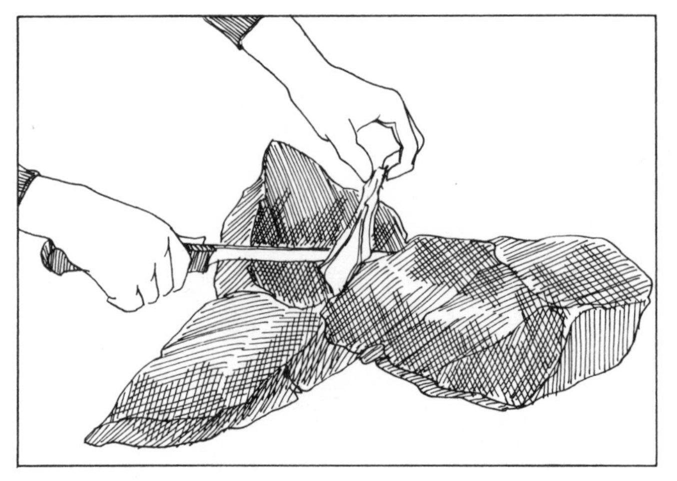

L-49. TAKING OUT LEG BONE

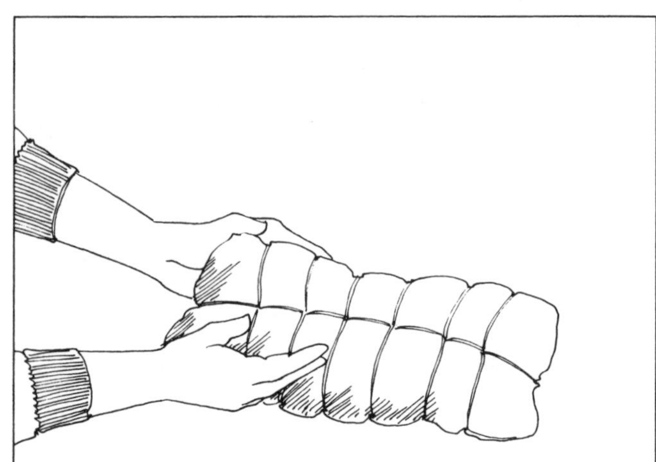

L-51. TYING UP BONELESS LEG

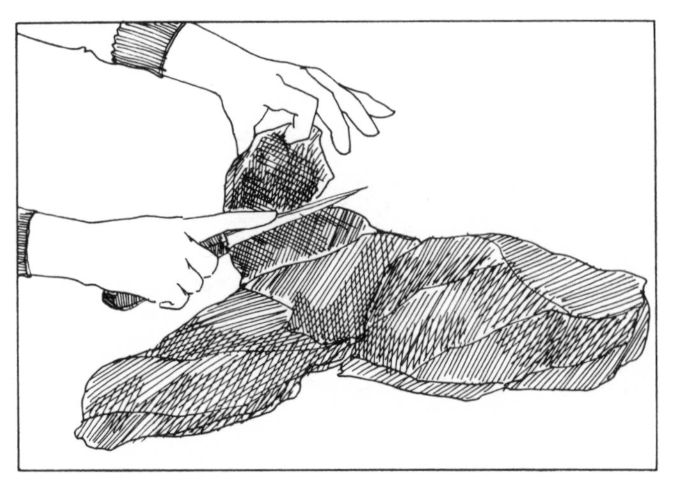

L-52. LEG SPREAD OUT FOR BUTTERFLYING

BUTTERFLIED LEG OF LAMB

Remove all the bones, fat, glands, etc., as in preparing the boneless leg. Spread the lamb out. Some parts will be thicker and higher than others. Holding the knife almost horizontally, cut into those thick parts to open them up as though one end were hinged. They can then be folded out flat to make the spread-out lamb more uniform in thickness (L–52).

Lay wax paper over the lamb and pound the lamb with the flat side of a cleaver until it is fairly flattened and faintly resembles the outspread wings of a butterfly. It is now ready to marinate or grill as your recipe dictates. The butterflied lamb can also be stuffed and rolled into a roast.

KABOBS

For kabobs it is not necessary to bone the leg; but trim all the fat away from that portion you intend to use for the kabobs. Kabobs can be cut from the shoulder section of the lamb, but we feel it is best, from the point of economy and flavor, to cut them from the back part of the leg. First find the membranes on the leg and cut around them for individual pieces. After you've cut the pieces off, simply cube them to make the kabobs (L–53).

L-53. CUBING PIECES FOR KABOBS

LAMB SCALOPPINE

The top round part of the leg (the biggest and roundest section) makes the best scaloppine. Cut the meat from the leg at this section, remove the fell, and trim away the fat.

Examine the meat itself and find the grain. Cut into reasonably thin slices *against* the grain for scaloppine. Then lay the scallops flat on a sheet of wax paper. Place a second sheet over the pieces and pound them to scaloppine thinness with the flat side of a cleaver.

4

Chicken

Today, poultry, particularly chicken, has come into its own. Certainly, chicken is one of the oldest and most versatile meats. The Sumerians and the ancient Egyptians enjoyed domestic chicken regularly, and they were force-feeding chickens to make tender and plump roasters long before the Strasbourg goose arrived on the gastronomic scene. For many years, American frontier cuisine dictated that chickens were only to be roasted, barbecued or stewed. The early settlers considered chickens mainly as a source of eggs, which had a multitude of uses. A hen was not considered ready for eating until she was past her egg-laying prime, and naturally the older birds were fit only for roasting or stewing.

All that has changed. Today's scientifically fed chickens are a high-quality food that is low in calories, high in vitamin content, and low in fat; much of the initial fat content tends to seep out during cooking. But age and weight are directly related in chickens, and the feeding of poultry governs the taste and quality of the chicken, even as it does with other meats.

As with other mass-produced commodities, however, marketing and sales considerations have affected the chickens produced for today's consumer. Some are injected with various stimulants to hasten growth and to add fat, allowing birds to be brought to the market more rapidly. And, though perhaps to a lesser degree than other meats, chickens have also been subject to the general increase in meat prices.

Knowing how to get the most out of your own chicken at home will save you considerable money, especially with regard to chicken parts, and will allow you the freedom of choice that is becoming more and more restricted in today's prepackaged marketing operations. Whether you purchase your whole chicken at a retail outlet or at a wholesale market or supplier's, certain factors will remain the same. Knowing what to look for is no less important in poultry than in any other meat.

Birds designated U.S. Grade A are those of the highest quality. They have more meat and have been fed quality grains. Grades B and C are seldom found marked as such in retail outlets, but numerous large retail outlets do market poultry that carries no grade. These birds are often Grades B or C. It is also possible for Grade A-marked birds to be bruised in handling, shipping and keeping *after* being graded. Your own knowledge and educated eye will be your best insurance, especially at the commercial or wholesale markets.

Look for a bird with a fleshy, firm breast, and rounded thighs—a Raquel Welch of a chicken. The skin should be on the thin side, not thickened with fat. A good chicken should be a pale yellow color, an indication that it has been fed with a scientifically prepared mix. A deep yellow color may indicate a feeding mixture that has products added to stimulate growth, resulting in more fat as well as meat. Avoid whitish chickens, scrawny birds with high and pointy breastbones, and those with dried-up skin. Also avoid chickens with bruise spots and discolorations. Finally, avoid chickens in plastic packages that contain pockets of water. This indicates that the birds have been frozen, thawed, and refrozen.

Sometimes, especially at large commercial suppliers, the origin of the chicken will be missing, and it is important that you know that every chicken, graded or ungraded, carries a number on its tag or wrapping. Often this number is inside on the wrapping around the neck and giblets. Should you want to know where your chicken came from, send this number to the U.S. Department of Agriculture and they will tell you exactly where your chicken originated.

The common designations of type for chickens can be misleading, especially as used in the retail markets, where terms such as "young chicken," "broiler," "fryer," "broiler-fryer," "roaster," "stewing chicken," "fowl," and "capon" are sometimes used carelessly. So that you may understand what you are purchasing, and even more important, what you need not concern yourself about, this is how the chicken market breaks down for all practical purposes:

Broilers: Very young chickens weighing from 1 to 2¾ pounds. Raised in special feeding pens, not allowed to run, they have a fine-textured, very white, delicately flavored meat.

Fryers: Slightly older, from 9 to 10 weeks of age, these birds are often let run free before being marketed.

Broiler-fryers: This is a catch-all phrase most often used by retail markets. In truth, most of these birds are suitable for either broiling or frying. However, this category often includes birds up to 3½ pounds. We do not recommend frying birds over 2¾ pounds. Birds heavier than this have too much meat for the heat to penetrate properly in the short frying period.

Roasters: From 2 to 3½ pounds, this is the traditional roasting chicken, with more meat than broiler-fryers.

Today's broiler-fryers can also be roasted; roasting time is shorter, because they weigh less than regular roasters. These three, often interchangeable classifications in today's market—broiler, fryer, roaster—are the three types of chickens you will most often find and buy in the market.

Fowls: Sometimes called stewing chickens or broiler hens, fowls are older, about 1½ years of age and from 3½ to 7 pounds in weight. Some people find their meat more flavorful than that of other chickens. They need long, slow, moist cooking, but they can be an economical buy when you want meat for dishes made with "cooked" chicken, such as casseroles, salads, chicken pies, and Tetrazzini.

Capons: These are male birds with their reproductive organs removed at an early age. They grow fatter faster and are usually shorter and plumper in conformation. They can be used for roasting and have a flavor somewhat sweeter than regular chicken.

Because of its smaller size, a chicken is naturally a good deal easier to handle than a side of beef. However, to cut a chicken properly requires deftness and artistry. The results will be worth your while.

SPLIT CHICKEN

Place the chicken on a flat surface breast down, with the chicken's tail facing toward you. Take hold of the chicken's tail with one hand and place the knife on the chicken so that the last quarter of the blade near the hilt is nearest the tail (C–1) and the tip of the blade just forward of the center of the chicken.

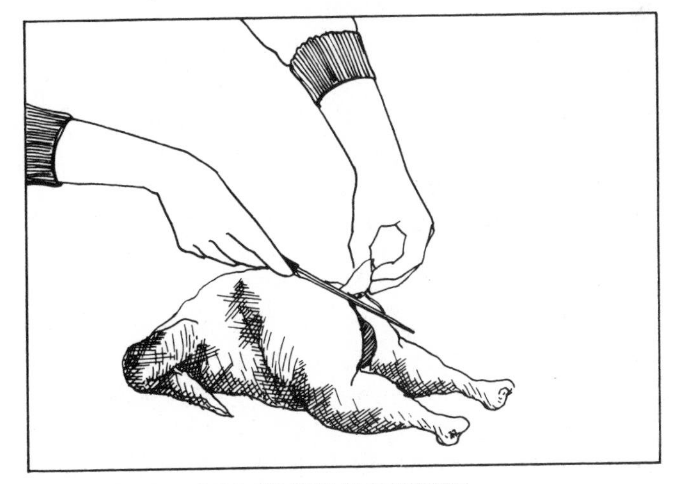

C–1. PREPARING TO SPLIT THE CHICKEN

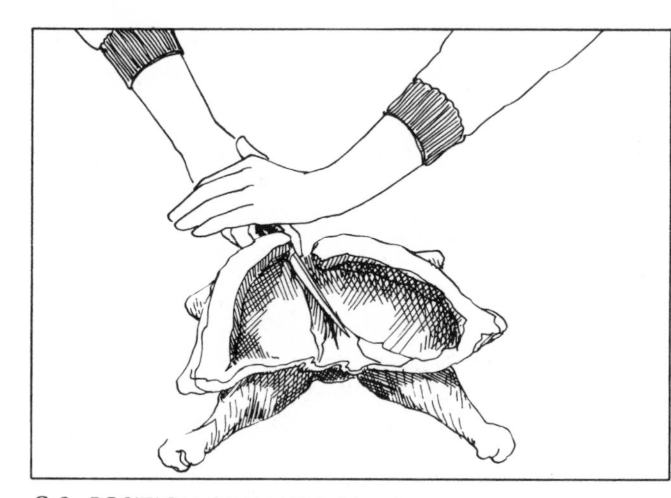
C–2. POSITION OF HANDS FOR CRACKING BREASTBONE

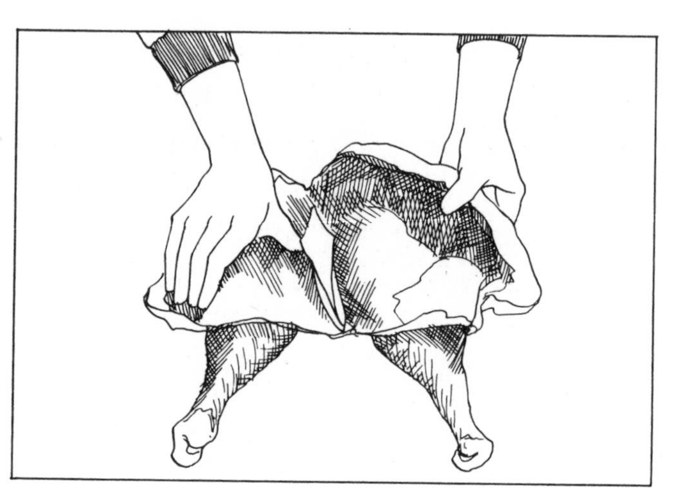
C–3. SPLIT CHICKEN

Pressing sharply with your wrist, make one snapping incision, cutting straight down and through the bird to cut the chicken almost in half. The breastbone will still be intact. Use the tip of the blade to crack the breastbone. Place the knife against the bone, half horizontally, and, using the palm of your hand, come down sharply on the handle of the knife (C–2), and the breastbone will crack open. You will have two halves, or a split chicken (C–3). The outer skin will still be connecting the two halves of the chicken. Turn the entire chicken over, roast as is (or broil) skin side down for three quarters of the whole cooking time, skin side up for remaining one quarter of broiling time.

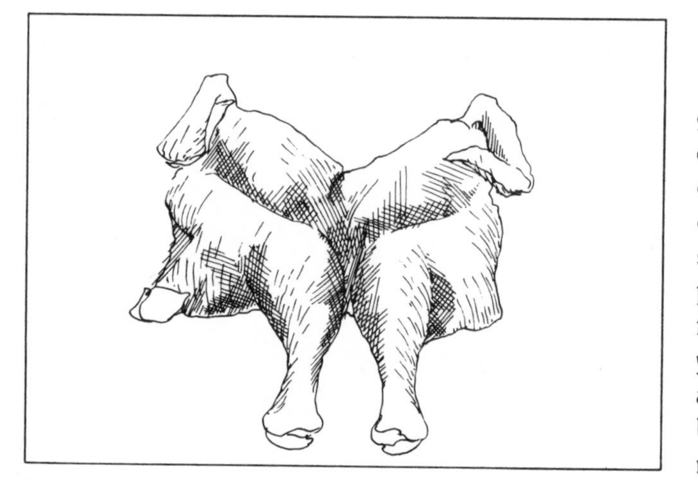
C–4. BUTTERFLIED CHICKEN

BUTTERFLIED CHICKEN

You may want to remove the breastbone altogether, for butterflied chicken. Once you have cracked it, this is not difficult. To crack, place thumbs on top of the breastbone, over the skin, your fingers on the underside, and, with a single quick motion, snap the breastbone. Keeping the hands in the same position, hold with the left hand and push the middle finger of your right hand under the breastbone. Run your right thumb down the length of the breastbone and the bone will come loose. Place the bird on its back and use a knife to cut off the remaining connection of breastbone. Using your right thumb and middle finger to sweep down the length of the breastbone leaves a clean separation without meat clinging to the bone (C–4).

QUARTERED CHICKEN

When you have cut the chicken in half (C–5) and have opened it, use thumb and forefinger to widen the center. Then use a knife to cut the chicken into two halves. Place the knife between the breastbone and the meat underneath and remove the breastbone (C–6). To quarter, place the knife along the line of the chicken's thigh and cut down and sever the thigh and leg from the remainder of the half chicken (C–7). Repeat with the other half; the chicken is now quartered.

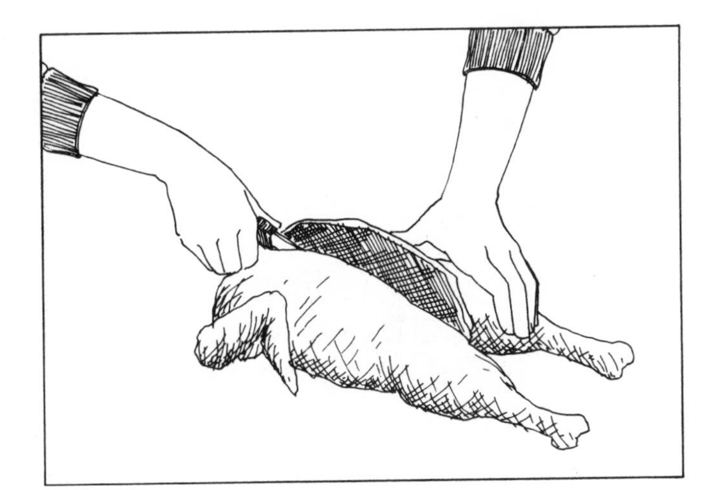

C-5. CUTTING CHICKEN IN HALF

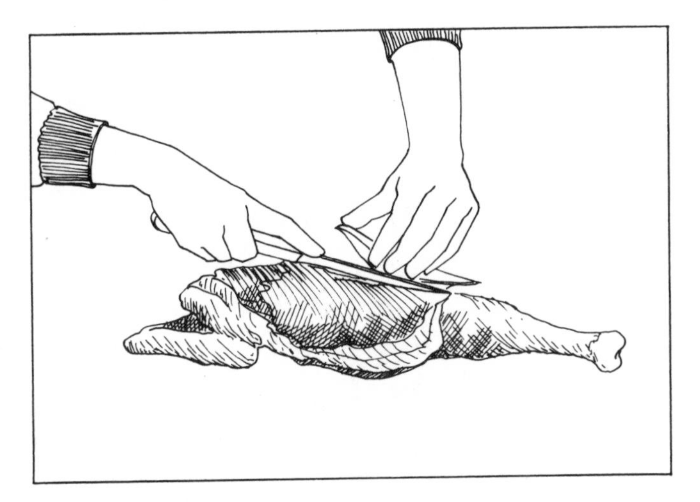

C-6. REMOVING BREASTBONE

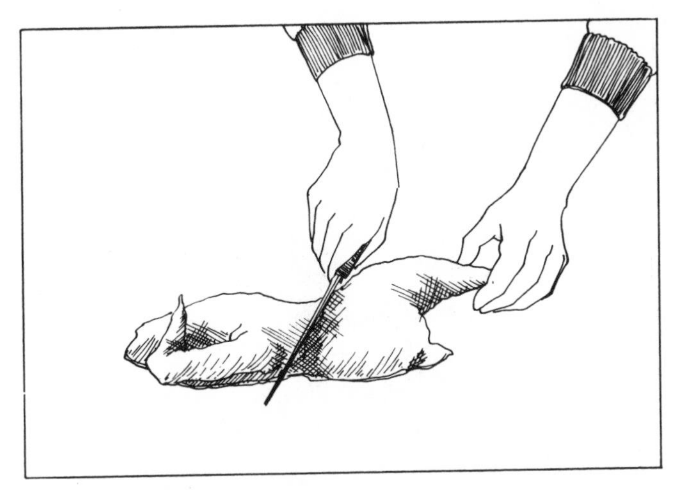

C-7. QUARTERING CHICKEN

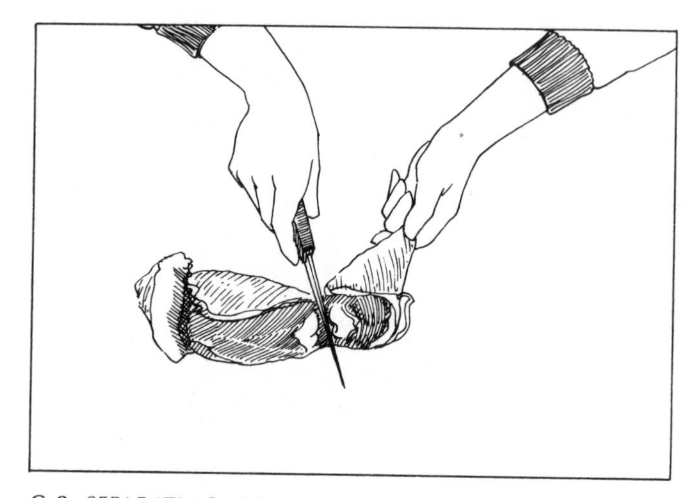

C-8. SEPARATING LEG AND THIGH FOR CUT-UP CHICKEN

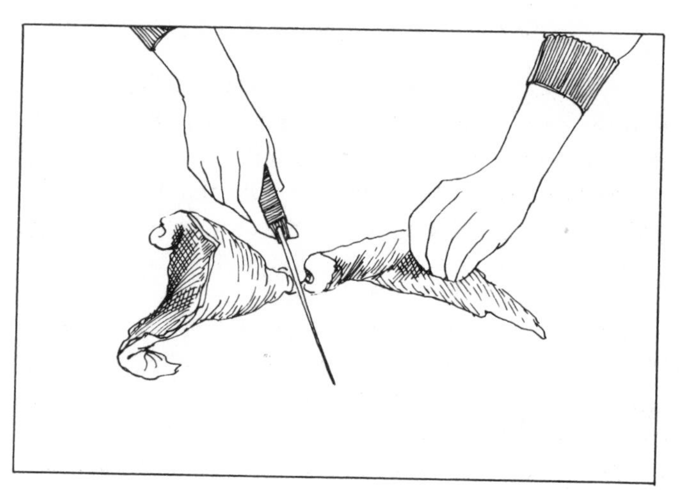

C-9. CUTTING WING FROM BREAST

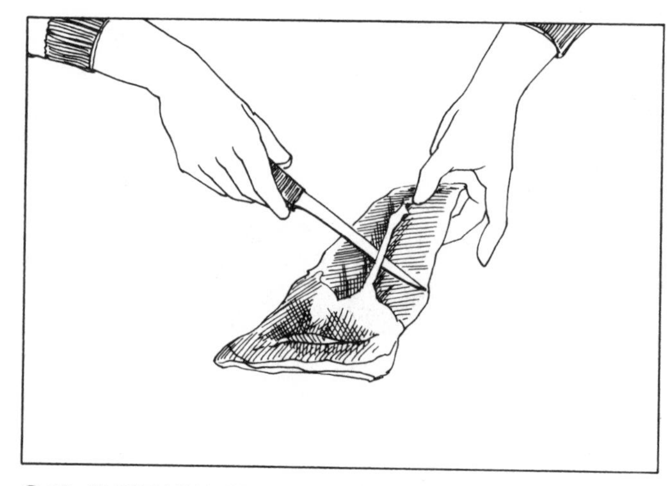

C-10. SLIPPING KNIFE UNDER THIN RIB BONE

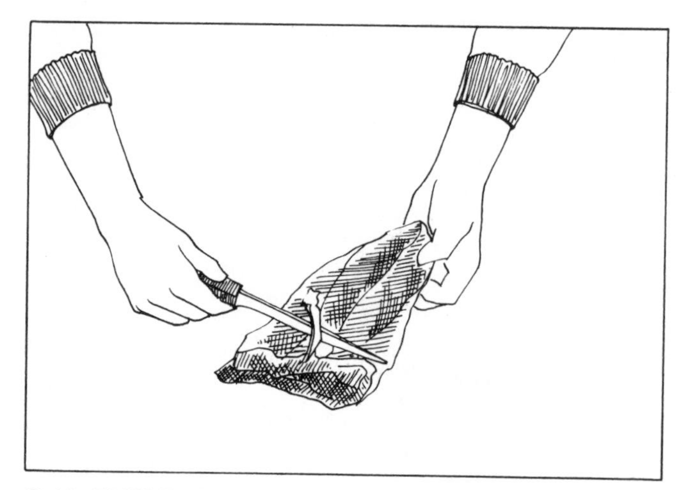

C-11. FINISHING FINAL CUT TO REMOVE BONE

CUT-UP CHICKEN

This is the chicken cut into eighths for serving pieces. After the chicken has been quartered (see above), turn the leg section so that the inside of the leg is up. You'll see a white cartilege covering the top end of the leg bone. Cut just above this to separate the leg bone from the thigh (C–8). Make the first cut and you will see the ends of both bones. Cut between them. Hold the wing away from the breast. Cut just above the top of the wing (C–9). Place the knife up against the large breast section and cut through with one thrust. The breast section is left intact.

BONELESS BREAST

Lay the breast skin down and you will see the thin rib bones. Slide the knife under these thin bones, starting at the top end, and cut carefully down, keeping the knife close to the underside of the bone, so you do not cut into the meat. Lift this thin section of the bone when you reach the base (C–10) and then cut into the tissue to remove the entire rib-bone section (C–11).

BONELESS STUFFED BREAST

Lay the boneless breast skin down and lift the top flap and you'll see a small pocket. Stuff lightly and roll or fasten. Boneless breast can also be cut into cubes for kabobs.

BONELESS CHICKEN CUTLETS

Cut through the natural opening on the side of the breast to prepare it for cutlets and/or scaloppine (C–12). Take the flap where you would place your stuffing and with a knife, carefully open the flap entirely to lengthen it. This is a chicken cutlet with the skin on. If you want the skin off, lay the breast flat with the skin up. Lift a corner of the skin on the thickest part of the breast and the separation of skin and meat will be visible. Move the knife with a gliding motion along the underside of the skin and tug gently at the skin as you do so (C–13). The skin will come off easily. Remove any excess yellow fat after the skin is off.

CHICKEN SCALOPPINE

Place the chicken cutlet between two sheets of wax paper and pound thin with the flat side of a cleaver. The chicken scaloppine can then be filled or rolled. It can also be cut into thin strips for use in Oriental dishes.

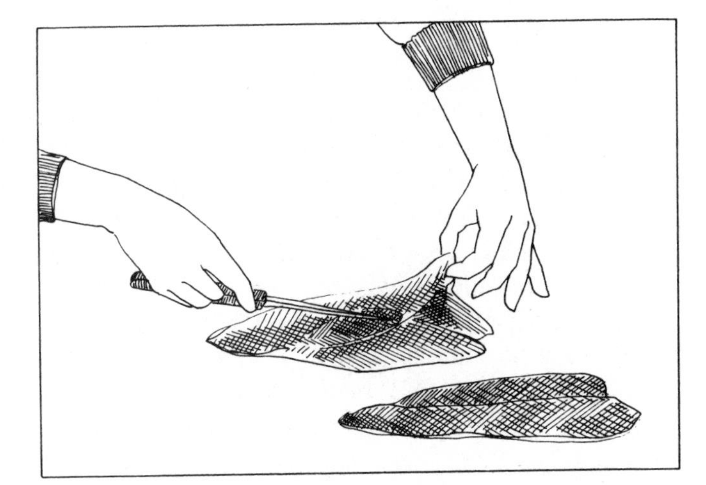

C–12. THE NATURAL OPENING FOR CUTLETS OR SCALOPPINE

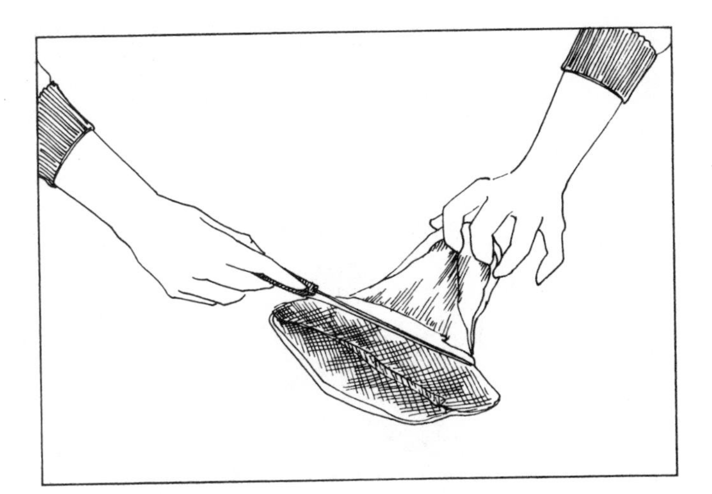

C–13. REMOVING SKIN

BONELESS DRUMSTICK

French the drumstick by stripping the tip of its skin and meat (C–14). Loosen area around the bone, using the tip of the knife (C–15). Take the end of the loosened bone in thumb and forefinger, as you would in pulling the cork out of a bottle of wine. Slide the fingers of the other hand down the bone, pushing the meat (C–16) in front of the fingers until it comes off. The meat will come off inside out (C–17). Pull the bone away entirely. With the fingertips, turn the meat back right side out on your finger and you have a boned chicken leg for stuffing or for hors d'oeuvre.

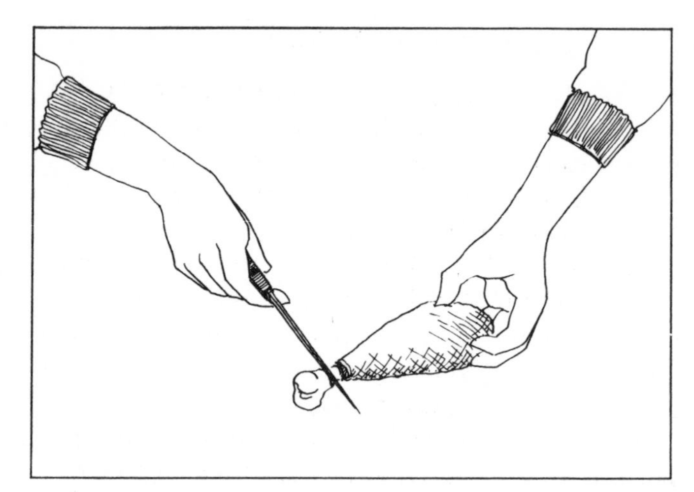

C–14. FRENCHING DRUMSTICK

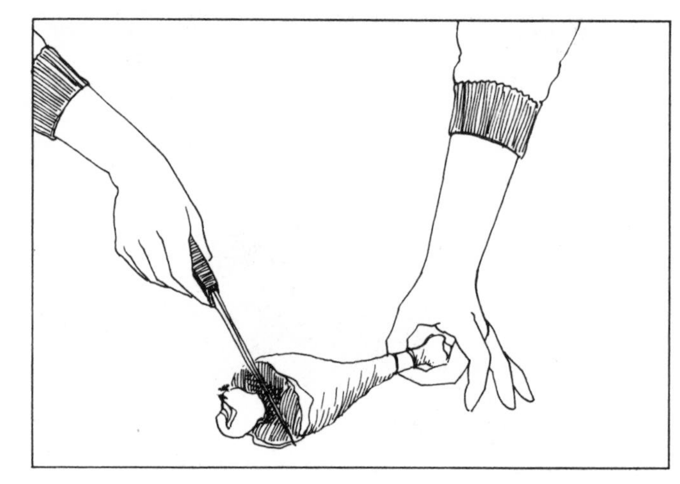

C–15. LOOSENING MEAT AROUND FRONT OF BONE FOR BONELESS DRUMSTICK

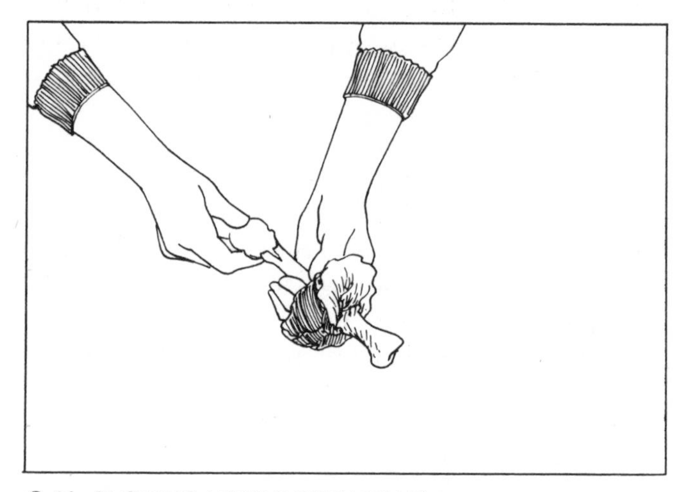

C–16. PUSHING MEAT DOWN BONE

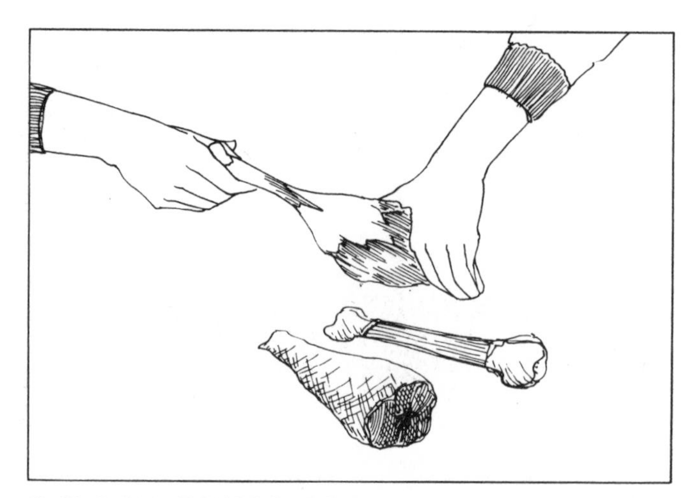

C–17. PULLING THE BONE OUT

CHICKEN DUET

This will take work on your part but it makes an elegant party dish, visually as well as gastronomically rewarding. A stuffing is required for this dish.

Remove the breastbone from half a chicken, along with the piece of wishbone. Take off rib cage as described earlier (C–18), with the other small bones (C–19). When the rib cage is removed, the half chicken still has its wing, drumstick, and thigh. When you remove the rib cage, take care to cut between it and the wing bone, which stays on.

Now remove the hipbone. Do this by snapping it out with your hands. With the knife, go underneath to the thigh bone, holding firmly to the chicken (C–20). Remove the thigh bone *but not* the leg bone (C–21). Lay half chicken out flat. On one end will be the leg and thigh (with the thigh bone removed) and on other end will be the wing. In the center will be the boned section.

Pull the wing outward to stretch the webbed skin area at the point of the wing joint. Make a slit in this webbed area with the tip of the knife (C–22). Then place the stuffing in the center of the chicken half. Fold the boned breast part over the stuffing, then bring up the other side to completely cover the stuffing. Raise the leg bone so that it extends into the air, cross the wing over it from the other side, push the leg bone through the slit you made in the webbed wing skin, and you have a Chicken Duet boned and ready for roasting (C–23). Roast for approximately 45 minutes in a 350° oven. Serve one to a person.

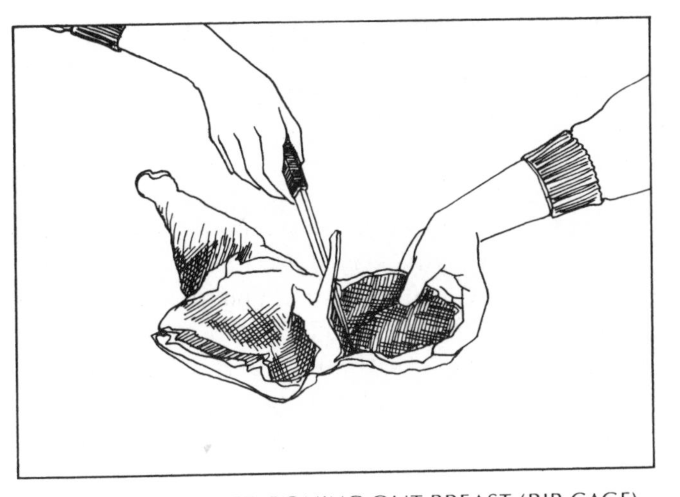

C-18. CHICKEN DUET: BONING OUT BREAST (RIB CAGE)

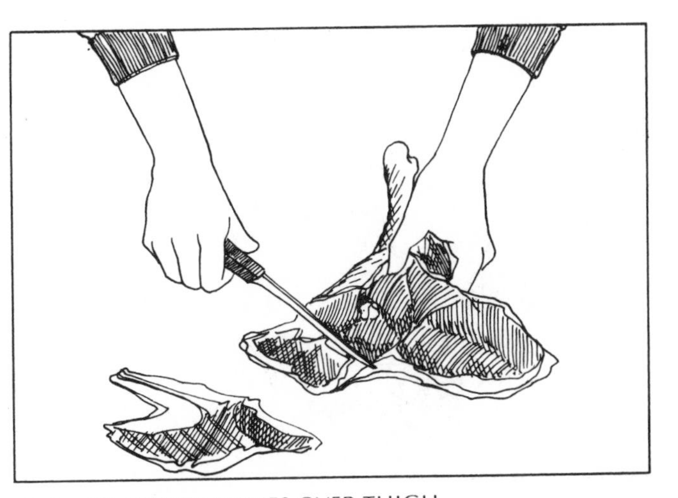

C-19. REMOVING BONES OVER THIGH

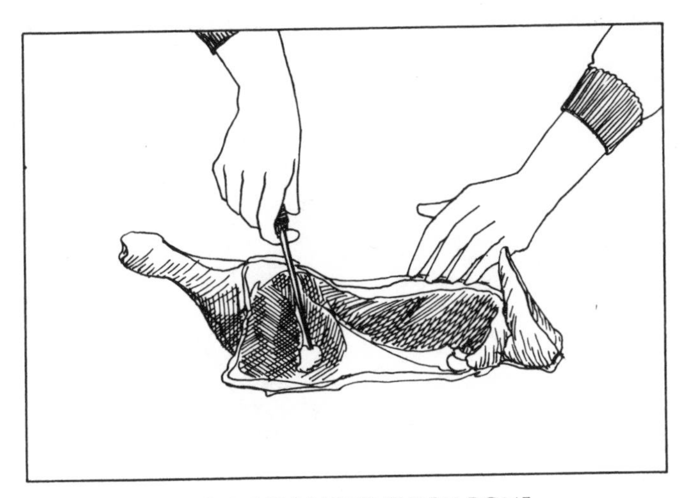

C-20. WORKING KNIFE UNDER THIGH BONE

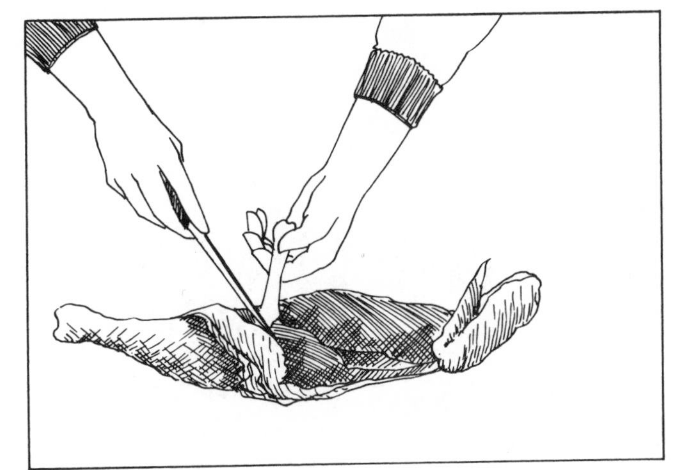

C-21. REMOVING THIGH BONE ONLY

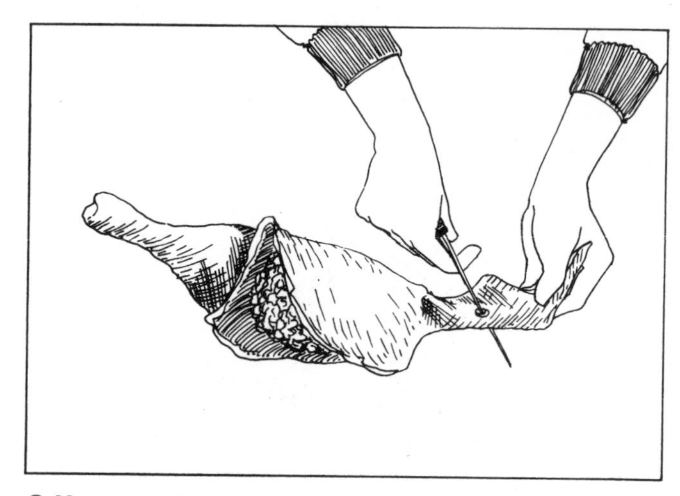

C–22. MAKING SLIT IN WEB OF SKIN

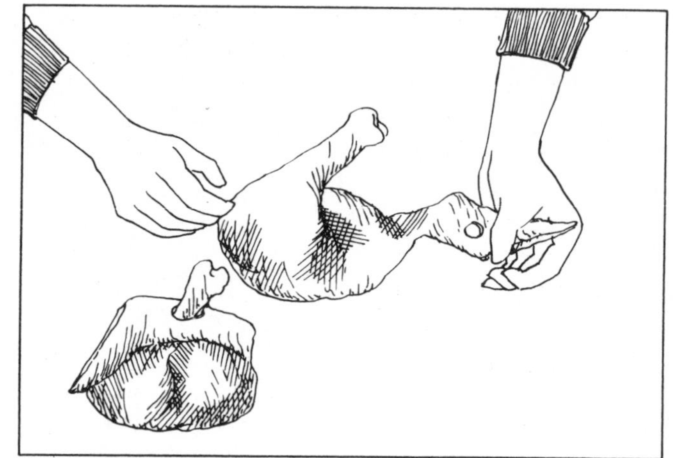

C–23. CHICKEN DUET WITH LEG THROUGH WING

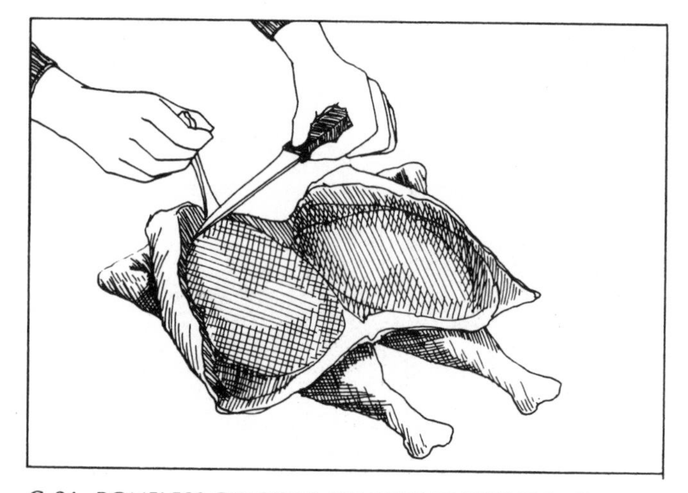

C–24. BONELESS CHICKEN: REMOVING WISHBONE

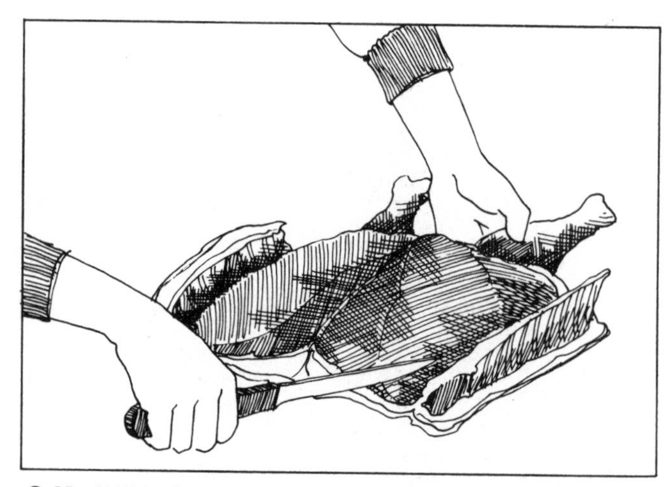

C–25. SLIDING KNIFE UNDER BONES OF RIB CAGE

WHOLE BONED CHICKEN

A whole, boned chicken is the basis for a special dish, stuffed in special ways or pinwheeled, where stuffing is put on top of flattened chicken and both are rolled into a cigar shape.

First place the chicken with the breast down and make the sweeping thrust with your knife as described in the section on split chicken. But do not sever the skin on the breast-down side. Spread chicken out as for butterflying. Remove fat and then take out breastbone. Use both thumbs on both sides and pull off by hand. With the knife, remove both halves of the wishbone (C–24). Turn the chicken around so that the legs now face you.

Slide your knife under the bones of the rib cage and work down to remove the entire rib cage. This should come out in one piece. Turn the chicken as you work to make it easier, and so that the knife blade is always gliding away from you as you cut (C–25).

When you've taken out both rib cages, two backbone sections still remain. Snap each with your hands, then cut around the bones with the tip of the knife blade and when loosened, pull the bones off (C–26). Remove the thigh bones as described in the earlier section on Chicken Duet (see C–27).

Lay chicken out flat again, with the legs toward you. Cut around the skin and meat at the end of each leg (C–28). Pushing the leg up with one hand, peel the skin back from the large inner joint of the leg. Working the tip of the knife around, free the joint and cut down along the bone to loosen all meat (C–29). You should be able to pull the leg bone out by hand (C–30). Push the meat back into the skin of the leg (C–31).

Once again, lay the chicken out flat. Begin to stuff the chicken by placing small amounts of stuffing into the legs where bones have been removed so that the legs are filled out to their proper shape. Place the main portion of stuffing on the flat center of the bird (C–32). Make a small mound of the stuffing. Fold the skin from each side to cover the stuffing until the bird resembles its original shape. If you see you need more stuffing, fold skin back and add it now. When the chicken has been properly stuffed, weave a trussing needle through the skin and tie ends together (C–33). Do the same with skin between the 2 legs.

Sew the center of the chicken, threading cord through the flaps of skin, and leave a foot of cord hanging (C–34). Turn bird over, put needle through tip of the leg, the tail, and tip of the other leg. Pull remaining foot of cord through and tie (C–35).

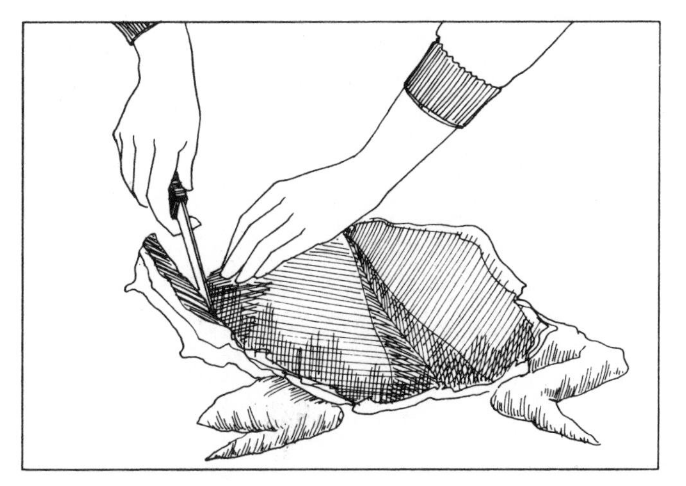

C-26. REMOVING BACKBONE

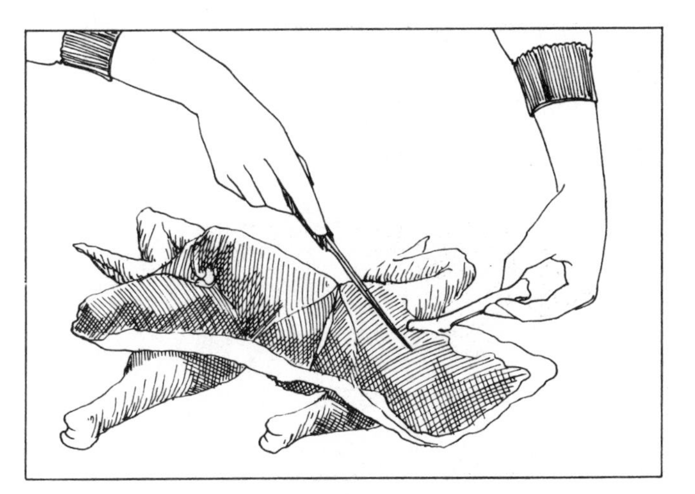

C-27. REMOVING THIGH BONES

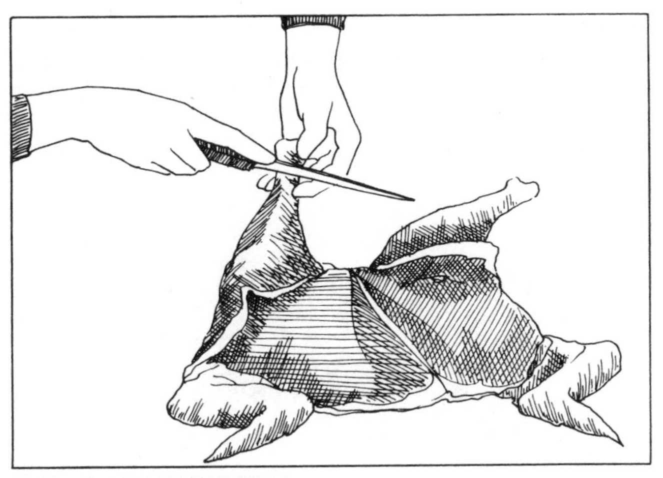

C-28. CUTTING LEG SKIN

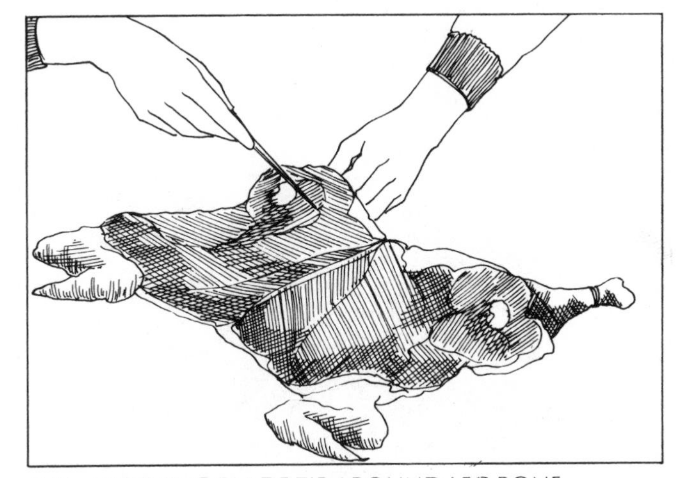

C-29. WORKING BLADE TIP AROUND LEG BONE

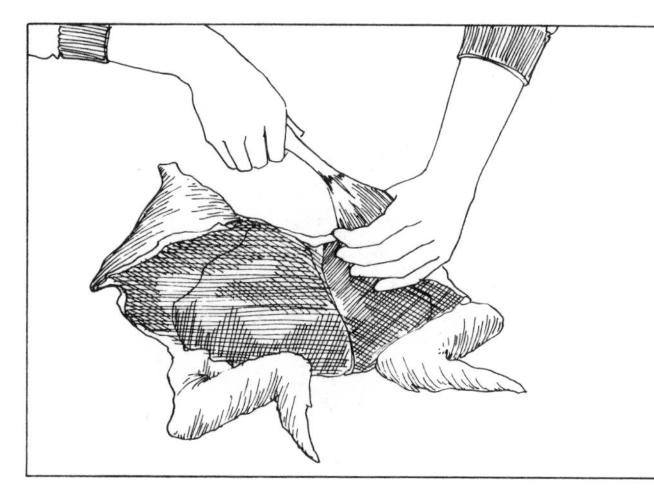

C-30. PULLING LEG BONE OUT

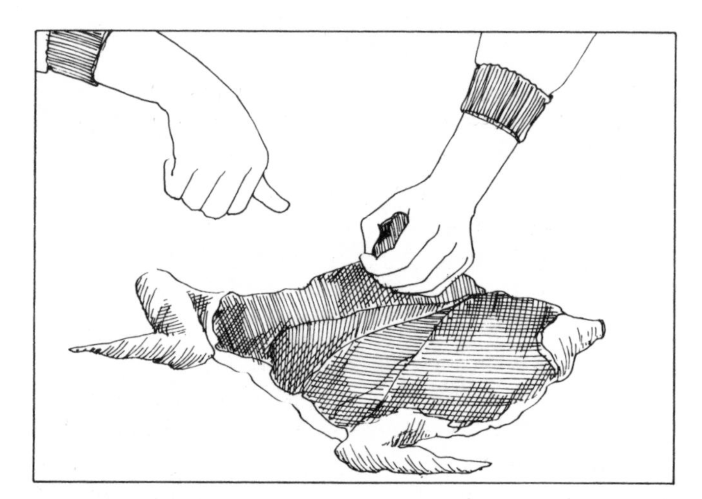

C-31. PUSHING MEAT BACK INTO LEG SKIN

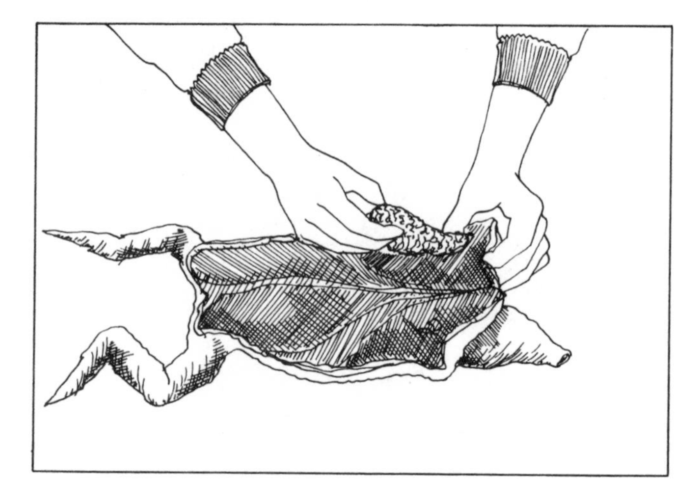

C-32. STUFFING LEGS

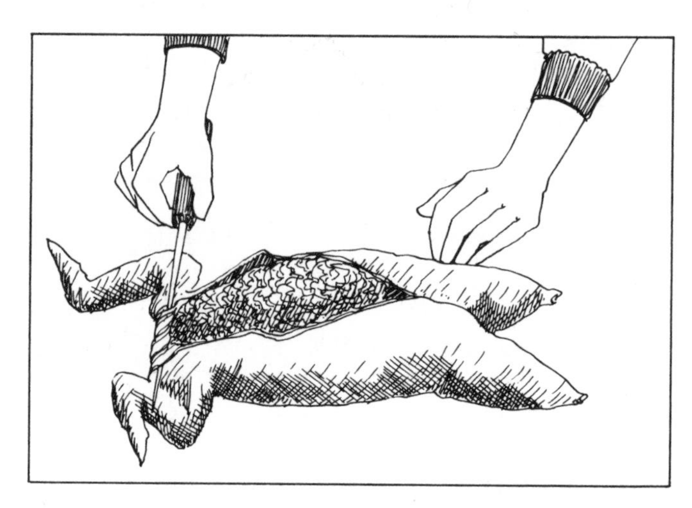

C-33. BEGINNING TO SEW WITH TRUSSING NEEDLE

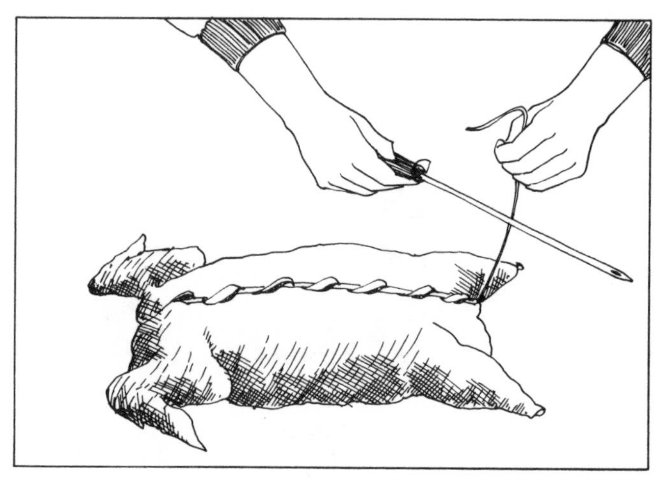

C-34. BACK OF CHICKEN SEWN UP

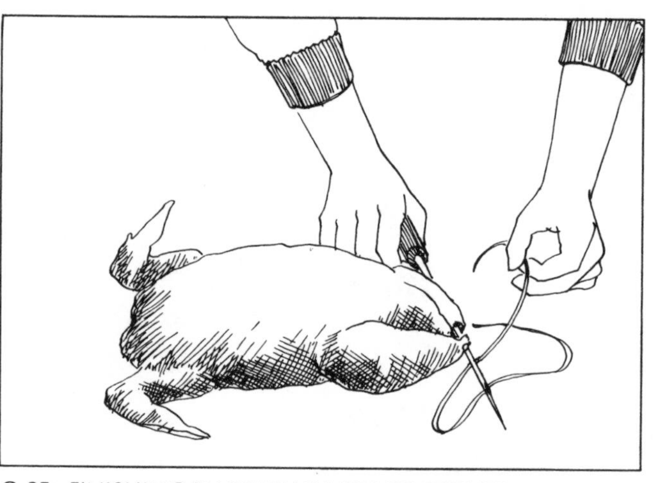

C-35. FINISHING BY SEWING LEGS TOGETHER

CHICKEN PARTS

If you like or need only certain parts of a chicken, and you purchase only these parts from your market, you will most likely use the parts pretty much as you've purchased them. However, if you've bought whole chicken breasts, you may want to halve them at home. To do this, place the breast skin side down so the inside of the breast is facing you. You'll see a white, soft cartilege that looks like a bone. With a knife, make an indentation on both sides of the cartilege just under the breastbone. Place a thumb on each side of the rib cage and press down, freeing the breastbone. Run thumbs down both sides of the breastbone (C–36). Take the top of the breastbone in your hand, holding the rest of the chicken down, and pull out the entire breastbone (C–37). Cut the chicken breast in half. For boning the other parts of a chicken, see previous descriptions.

The procedure for cutting or boning a Rock Cornish Hen, a large chicken, a capon, or a turkey is the same as for ordinary chickens, as described above. The very small birds and the very large are, of course, more difficult. We suggest practice on medium-size chickens first.

Two notes to bear in mind: Do not discard any bones. Use them for stock, broth, soups, etc. Giblets should always be stored separately, as they have a shorter refrigerator life than the rest of the bird. (See the chapter on freezing.)

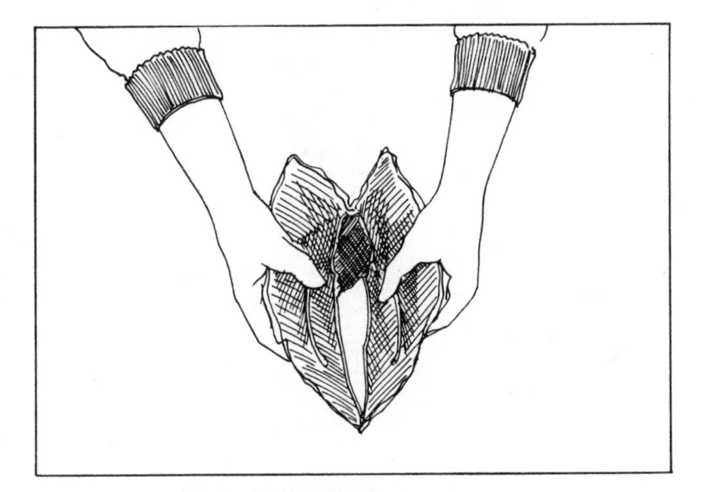

C-36. WORKING OUT THE BREASTBONE

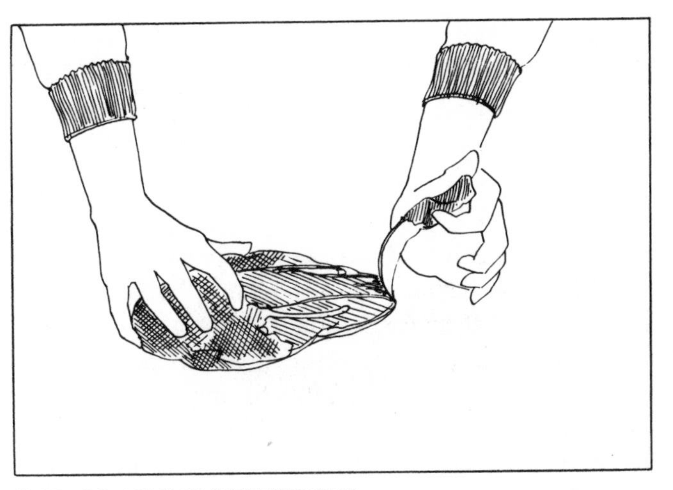

C-37. REMOVING BREASTBONE

Most game birds, in which we include duck and goose, are roasted whole after preparing according to recipe. The fancy boning of game birds is rarely done, being called for in only relatively few recipes. Moreover, it requires a degree of advanced skill that goes beyond that required for the other home carving explained in this volume. Therefore we do not recommend the intricate boning of game birds until the skills needed for the other procedures herein are thoroughly mastered. Occasionally, however, a recipe calls for quartering. To quarter game birds simply follow the same procedures given for quartering a chicken.

It is useful for you to know the varieties of such birds generally available, whether you purchase them at your market or receive them directly from a hunter. Game birds vary tremendously in taste, texture, and quality. Individual preferences will eventually decide your choices, but a certain amount of exploration is both necessary and enjoyable.

It is important for you to keep in mind that most ducks and geese sold in the marketplace are not truly game birds. Most are raised domestically on large farms, fed controlled diets, and prepared for sale in much the same way as is other domestic poultry. The taste of these birds may be more appealing to the average palate, but you should be aware that there is a vast difference in taste between these game birds and the wild variety. Eating wild duck after being used to commercially marketed duck can be an unexpected experience.

In general, it may be said that domestically raised and fed ducks and geese are less robust in flavor than the truly wild birds, lighter and blander, often less stringy and often meatier. Wild ducks and geese are stronger in flavor, darker, and much more defined in taste. If there is one area where the saying "You are what you eat" is valid, it is in this area of game birds, whose flavor reflects their diet.

Following are the kinds of ducks and geese you will most often find sold in today's marketplace, both wholesale and retail.

DUCK (DOMESTIC)

The major varieties on the domestic market are the White Pekin, the Mallard, and the Muscovy. Numerous breeders supply these ducks to the marketplace, and in time your own judgment will decide your preference for a particular supplier's product.

White Pekin: From 4½ to 5½ pounds, the White Pekin, as its name suggests, originated in China centuries ago. It is a large duck, fatter and juicier than most, and has a somewhat mild flavor. White Pekins are the ducks most often referred to as Long Island ducklings, though fine White Pekins are marketed by breeders from Virginia, Indiana, and Pennsylvania. Because there are so many suppliers of White Pekins, the quality of these birds can vary considerably.

Muscovy: From 2½ to 3½ pounds for hens and 4 to 7 pounds for drakes, the Muscovy originally hailed from the jungles of South America. It is raised and marketed, here, for those who want a bird with a somewhat more distinctively duck flavor. Though generally smaller than the White Pekin, it yields more breast meat.

Mallard: From 1½ to 3 pounds, the Mallard is the domesticated descendant of the wild Mallards native to the North American flyways. It is the smallest of the domestically raised ducks and also the most distinctively flavored, with rich, dark meat.

Mullard: This duck is a hybrid entry into the marketplace, a cross between a Mallard and a Muscovy. It is larger than both, and often larger than the White Pekin. It is a very meaty bird, juicy with a distinct, rich flavor. More often used in Europe, particularly France, the Mullard, like many crossbreeds, cannot reproduce itself and is therefore considerably more expensive than most other varieties of domestic duck. The Mullard weighs between 5¼ and 6½ pounds.

Most domestic ducks are marketed at ages between 2 months and 4½ months, so you, the buyer, have few worries about age. Yet from time to time older birds do reach the marketplace, and if you are obtaining a wild duck the age factor becomes highly important. Young ducks have a soft, pliable beak, while an old bird's beak is hard and inflexible. The legs of a young duck will be fairly smooth, an old bird's wrinkled and leathery. A young duck's webbing between the toes will be soft and easily torn, while an old bird will have tough, resistant webbing. Old ducks do not make very tasty eating.

WILD DUCK

Wild ducks are very different from the domestically raised variety. Their meat, even if they haven't dined primarily on fish, is of a different taste and nature than domestic duck. If they have dined long and heavily on fish, wild ducks may taste more of fish than duck. The best flavor is found in those ducks that normally inhabit certain areas for their food—grain fields, shallow water, and marshlands—and in those that are primarily surface eaters. The diet of these ducks contains enough grain, grasses, wild rice, corn, and sorghum to provide the desirable flavor and texture. There are too many varieties of wild duck to catalog here, but these are the ones most often used for the dinner table.

Mallard: 2 to 2½ pounds; dark, very rich meat, fine-grained in texture; strong flavored.

Pintail: 2 to 3 pounds; similar in taste to the Mallard; fine-grained meat, very rich and dark.

Canvasback: 2¾ to 3¼ pounds; less fine-grained meat, very rich and very dark; strong in flavor.

Black: 2 to 2½ pounds; very rich, dark, fine-grained meat; this duck is a cousin to the Mallard.

Redhead: 2¼ to 3 pounds; strong meat, often very fishy in taste, as this is a diving duck; dark and medium-grained meat.

Wood duck: 2½ to 3¼ pounds; dark meat, but less strong than many ducks; fine-grained meat; feeds on more seed, acorn, and fruits than most ducks.

GEESE

Like ducks, geese raised domestically on farms taste very different from the wild variety. The production of domestically raised geese is still a fairly restricted and spotty affair, and geese can be almost unobtainable during many months of the year. Most of the domestic geese on the market come from the Minnesota–South Dakota area. The collective farms of the Hutterites have developed the raising of domestic geese much as domestic ducks are raised. However, their marketing techniques do not at all match the commercial duck producers', and so domestic geese are often difficult to find.

Domestic geese, when available, are principally of three varieties: the White Embden, the Gray Toulouse, and the White China. None of these are related to the wild geese of North America, and all were developed from the European Graylag goose. Most are marketed between 2 and 6 months of age.

White Embden: This is the most popular of the domestic breeds, a large bird with a solid brick-shaped body. Firm, flavorful meat with plenty of breast meat. Ganders weigh approximately 15 pounds; hens 12½ pounds.

Gray Toulouse: About the same size as the White Embden, this is a deeper-bodied goose with perhaps a little more leg meat. Its body is shaped slightly differently from the White Embden but it weighs about the same. Flavorful meat, much the same in taste as the White Embden.

White China: The smallest of the domestically raised geese, with a longer neck than the other two. A more active bird, the meat is flavorful, perhaps a little firmer and stronger. Ganders weigh in at about 11 pounds and hens at about 8 pounds.

The meat of the domestic goose is stronger than that of domestic turkey, with a more distinct flavor, but not so strong as to jar the palates of those accustomed to turkey and domestic duck. The wild goose is very different.

WILD GEESE

Wild geese are not usually as fishy in taste as some wild ducks, but the flavor is generally far more robust and gamier. The most common wild geese are the Brant, Canada, and Snow Goose.

Brant: The smallest of the three species, the Brant has dark meat, very rich and very gamey. It sometimes has the fishiest-tasting meat of the three geese because of its feeding habits. Brants usually weigh from 4 to 6½ pounds.

Canada goose: This is the wild goose you most often see winging down the North American skies, honking as it goes and when it lands. The meat is dark, very rich, and strong, intensely flavorful, with medium texture; the breast meat is usually good. Canada geese weigh from 8 to 14 pounds.

Snow goose: Sheer white, except for the tip of its wings, the snow goose is a subarctic bird with lean meat, very rich and dark in taste, and very similar to that of the Canada goose. Snow geese weigh from 7 to 14 pounds.

There are other game birds, many of which find their way into recipes, particularly in those recipes of Mediterranean origin. So that you may have some idea of the taste and the size of these birds, we offer a brief look at their individual characteristics.

Partridge: Approximately 1 pound, partridges have white meat, reminiscent of chicken, but a fraction less bland. Obviously there is not much meat on a single partridge.

Pheasant: A cock pheasant weighs 3 to 5½ pounds, a hen 2½ to 3¼ pounds. The pheasant has white breast meat and dark leg meat. The meat is fine in texture and quite definite in flavor. The meat is firm and plentiful.

Quail: From 6 to 8 ounces in weight with very delicately flavored white meat, yet more distinctive than chicken. Very little real meat on each bird, and usually served at least in pairs, often in threes or fours.

Snipe: From 6 to 8 ounces in weight with rather dark meat. Flesh has a distinctive flavor and is fine-grained. Not a lot of meat on each bird.

Woodcock: From 10 to 12 ounces in weight, the woodcock has medium-dark, rich but lean meat. Reversing the usual, the leg meat is white.

Grouse: From 1¼ to 2 pounds, grouse has very rich meat, medium-dark in color and more aromatic than most game birds.

Pigeon (squab): Approximately 1 pound in weight, the pigeon has tasty meat, not very firm in texture, medium-dark in color, and faintly reminiscent of the dark meat of a domestic turkey, but more delicate.

Wild turkey: From 10 to 16 pounds, the wild turkey is very different in taste from the domestic turkey. The meat is richer, stronger, firmer in texture, and leaner and medium-dark throughout.

If you yourself hunt, or receive wild game birds from a hunter, it is extremely important that the birds be eviscerated as quickly as possible. With a sharp knife, slit the skin of the neck from the breast toward the head. Pull out and remove the windpipe and the gullet. Then make another slit from the rear of the bird, at the tail, back to the abdominal cavity large enough to permit withdrawing the entrails. *Note:* Take care not to break the gallbladder (attached to the liver), or the liver and giblets will be inedible. When the entrails are removed, clean the abdominal cavity with a damp cloth.

Plucking: Feathered wild game birds should not be scalded for plucking. They should be dry-plucked, the feathers pulled downward in the same direction as they grow. Pulling up tears the skin. *Note:* If you are hanging birds for ripening, do not pluck; hang them feathered until ready to use and then pluck. Pluck feathers using your thumb and forefinger. Pluck starting at the base of the neck and move in rows for the neatest effect. Ducks and geese have thick, heavy feathers, and paraffin may be used to help pluck them. Melt enough paraffin in a pot large enough to accommodate the duck or goose. Place a pail of cold water at your side. Remove the large wing feathers by hand, then immerse the bird in the paraffin. Remove it quickly and plunge immediately into the cold water. The wax will congeal and can be pulled off in large chunks, bringing the feathers with it.

Note: All fowl have an oil sac at the base of the tail. Never cook a duck or goose without first removing this triangular section of flesh. Do this after plucking, when the sac is clearly visible.

6

Pork

Pork is one of the most nutritious of meats. Perhaps that is why it has been a universal favorite for uncounted centuries. Pork is no higher in calories than other meats and is the leading food source of vitamin B_1 (thiamine). It is also one of the most digestible of foods, rated 96 to 98 percent digestible in government studies.

The ancient Chinese, 5,000 years before Christ, had developed the technique of smoking and salting pork. When the Tartars swept across Central Europe they brought this knowledge with them, and the Gauls became the first Europeans to make ham more or less as we know it today. Pigs accompanied the earliest settlers in America, and these hardy animals were able to thrive and forage on almost any kind of land and to endure the terrible winters.

Today, breeders produce hogs with leaner meat, and the quality of feed makes for sweeter taste. Nowadays, hogs are brought to the market in 3 major categories: 1. suckling pigs, usually a special holiday item, and they must be ordered in advance (these are piglets 3 to 4 weeks old that have been fed only on mother's milk); 2. young hogs, often called pigs, from 4 to 6 months old; 3. hogs, mature pigs usually from 9 to 14 months of age.

The most economical way to purchase pork is to purchase an entire pig. However, we suggest that in doing this, you purchase only young hogs (pigs) weighing approximately 60 to 75 pounds. These are the sweetest to the taste, but, more to the point, we feel they are a size that can be handled reasonably well at home. Look for a pig that has a firm outer skin, somewhat yellow-white in color, and tight, without excessive looseness to the touch. The skin should be smooth, free of wrinkles and hair roots. The bones should have some red (blood) in them, signifying a young animal. The lean meat should be a deep rose-pink and uniform in color, or with only slight gradations of color. Stay away from pigs with bruised spots, broken or obviously dislocated bones, or any malformations in the joints.

At the market, you may purchase an entire pig, half a pig, a whole loin section or half a loin section, the rear or shank quarter, or the fore- or shoulder quarter. To cut individual pieces from these sections, follow the procedures given below for those portions. We begin, however, with the assumption that you have purchased an entire pig.

There are 3 initial steps for preparing your pig for further sectioning. First, lay the pig on its side. To remove the head section, find the neck and shoulder and, cutting just above the shoulder and below the ears, make your first incision by knife, cutting straight across the pig. Then, using the saw, cut off the entire head section (P–1). Second, remove the feet. At the forefeet, find the joint and take off both feet at the joint. Use a knife to do this, cutting between the joint bones and the leg bones. For the hind feet, find the hock. You'll need your saw here to cut through and remove the hind feet. As you saw, bend the hind foot down to facilitate sawing. Remove both hind feet and save with the two forefeet.

Third, turn the entire pig on its back, facing you. Your pig will have come to you opened, the intestines, etc., cleaned out. With the opening running the length of the pig toward you, pull both halves of the skin on the sides back. Place the saw inside the cavity, starting at the front end, against the center of the backbone, which will be visible. Begin sawing from the front. This is a thick bone—the older the pig, the thicker the bone. Saw down the entire length of the pig along the backbone (P–2) until both halves come apart. You'll now have two halves with one foreleg and one hind leg each.

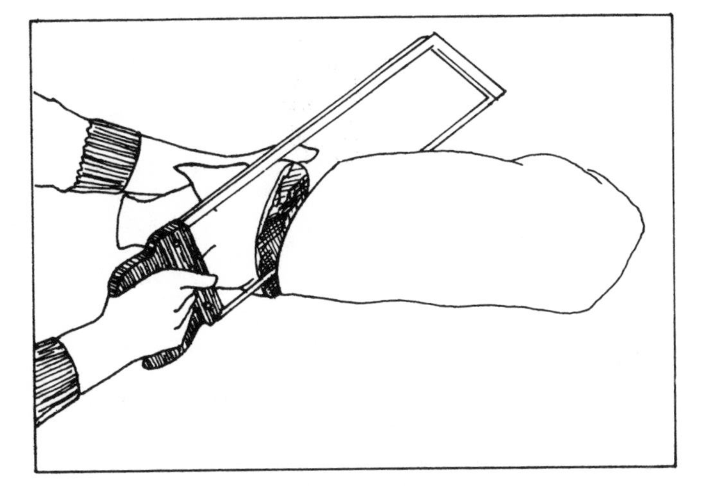

P–1. CUTTING HEAD SECTION OFF WHOLE PIG

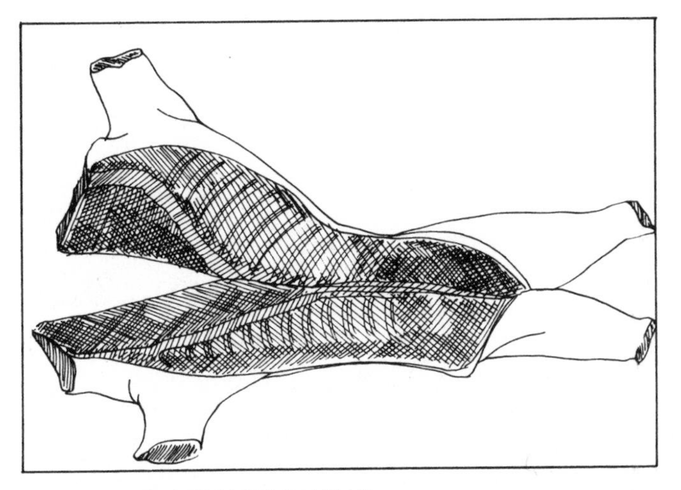

P–2. CUT LENGTHWISE IN HALF

SHOULDER AND BUTT

If your recipe calls for the following, cut them from the shoulder and butt sections:

Boston Butt (also called Pork Shoulder Blade Boston Roast; Pork Shoulder; Fresh Pork Butt)
Picnic Shoulder (also called Pork Shoulder Arm Picnic; Whole Fresh Picnic; Pork Shoulder Picnic)
Pig's Feet

LEG

If your recipe calls for the following, cut these pieces from the leg:

Hind Leg Whole (also called Pork Leg Whole; Fresh Ham)
Pork Leg Shank (also called Fresh Ham, Shank Leg Portion)
Sirloin
Hipbone Chops
Sirloin Chops

RIB AND LOIN

If your recipe calls for these pieces, cut them from the rib and loin section:

Loin of Pork/Center Rib Roast (also called Pork Loin Roast; Pork Loin Rib Half)

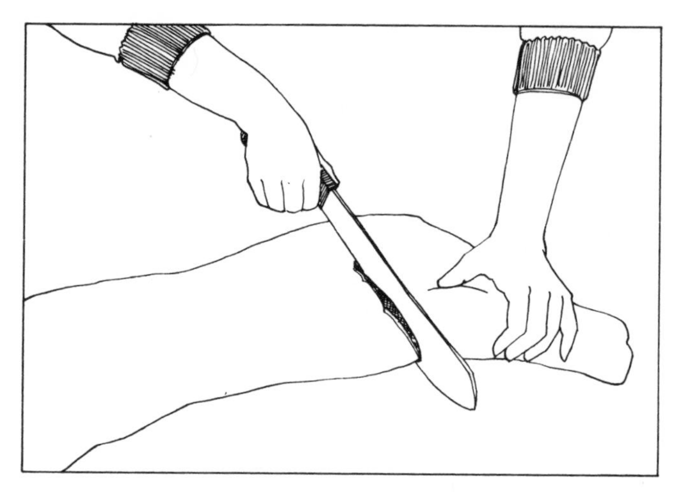

P-3. CUTTING THE FORELEG

Boneless Rib Roast (also called Boneless Pork Roast)
Center Cut Chops (also called Top Loin Chops)
Crown Roast
Loin Chops (also called Center Cut Loin Chops; Strip Chops)
Double Pork Loin Chops
Stuffed Pork Loin Chops
Butterflied Chops
Spare Ribs
Pork Scaloppine
Butterflied Chops
Ground Pork

SHOULDER AND BUTT

With a large (12-inch) knife, make a cut inward at the elbow of the foreleg (P–3). Then use the saw to complete the cut and free the leg. Move the shoulder up and down until you find the joint, and cut in at right angle. You will now have 2 cuts forming an "L" (P–4). This has removed the foreleg and given you the picnic shoulder. You will see the round shoulder bone in the cut section.

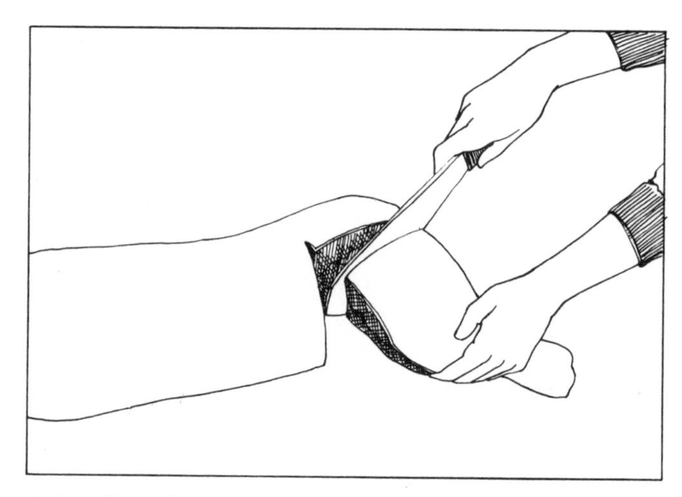

P-4. L-SHAPED CUT TO REMOVE FORELEG

BOSTON BUTT

Using the saw, continue on from the first incision you made in the picnic shoulder section to separate this entire piece (P–5). You will have a square piece, thick on 2 sides and tapering down somewhat toward the ends, called the Boston butt.

Turn the Boston butt with the skin side down. Place the saw between the bones in the piece and saw halfway down, until the bone is cracked. Make approximately 5 cracks (P–6). Now turn the entire piece over again and make 2 incisions with the knife for your saw, then saw down just enough to open the top of the blade bone, approximately 3 inches (P–7). Score the skin in a diamond pattern (P–8), tie along bottom bones, and tie again on top. The piece is now ready for the oven.

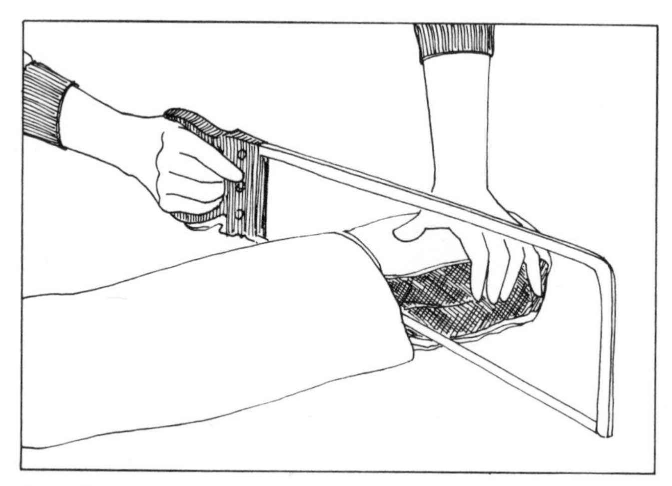

P-5. CUTTING THE BOSTON BUTT

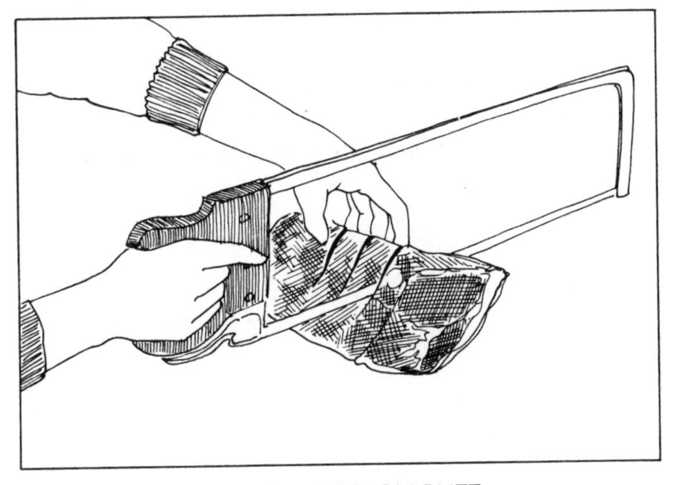

P-6. CRACKING BONES IN BOSTON BUTT

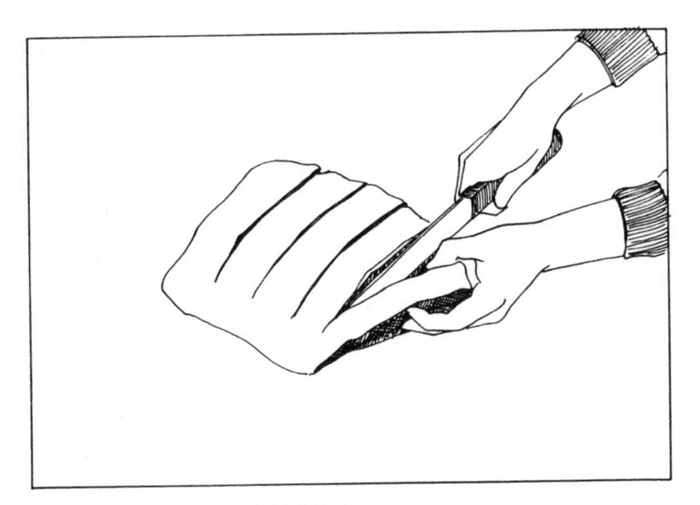

P-7. CRACKING TOP BONES

LEG

HIND LEG (WITH SIRLOIN ATTACHED)

First separate the leg from the body. Run your thumb along the back of the pig until you feel the hipbone. Using a saw, cut *on* the hipbone (P–9), cutting through the entire bone. This will give you the hind leg, which can be roasted whole.

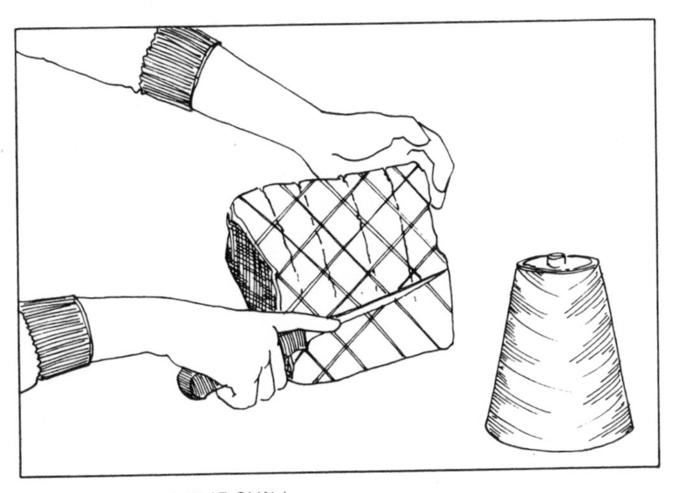

P-8. SCORING THE SKIN

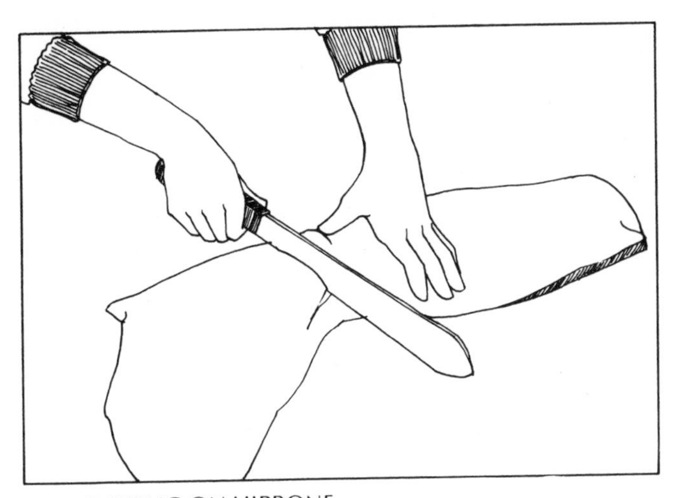

P-9. CUTTING ON HIPBONE

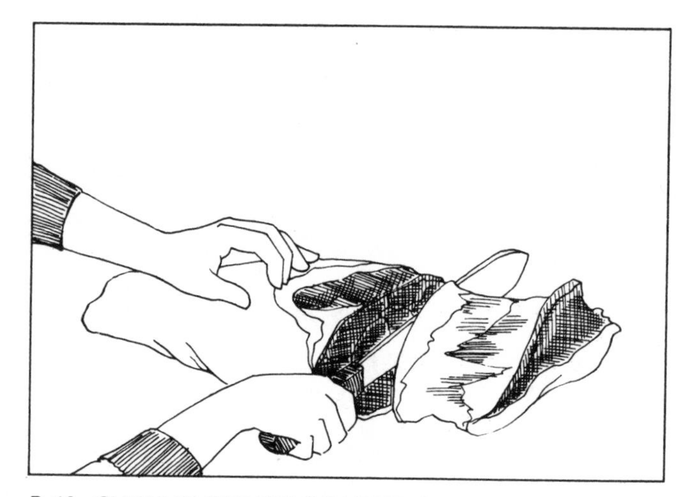

P-10. CUTTING OFF SIRLOIN SECTION

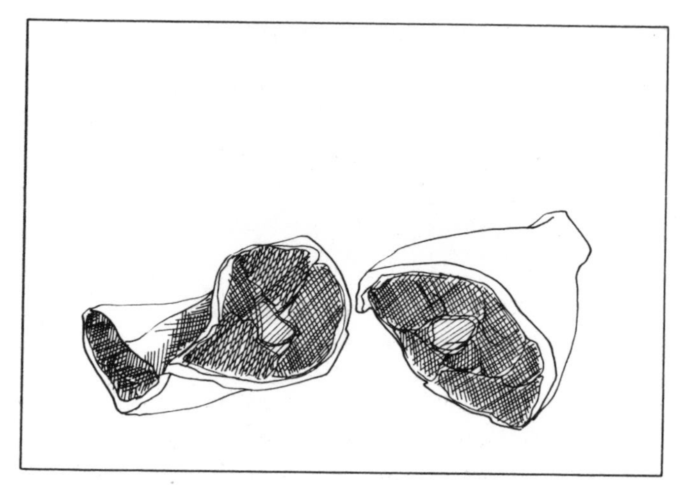

P-11. SIRLOIN AND HAM SECTIONS

FRESH HAM

Turn the hind leg on its back, skin side down. First cut off the sirloin section (P–10). A bone juts up in the middle of the section. This is the other end of the hipbone. Put your knife against this bone and cut downward, opening the leg into 2 pieces. Finish by using the saw to saw through. You will now have 2 sections. One will be the ham and the other the sirloin (P–11). *Note:* When you buy a fresh ham in the market, it will usually already be sectioned as explained above. This ham section may be smoked, roasted, or boiled, according to preference. Roasting is the method used most often. Score the skin in a diamond pattern before roasting.

SIRLOIN SECTION

The sirloin section may be cooked in its entirety. Simply place it in the oven following your own recipe.

HIPBONE OR SIRLOIN CHOPS

Slice these chops by cutting down the sirloin piece according to the thickness desired (P–12). Slice using a knife and cut through the bone with a saw. *Note:* These are *not* what are commonly called loin chops. They are of good quality but somewhat bonier than regular loin chops. Out of this section you will have 3 types of chops: flat, hip, and round-bone chops. You will be left with the end section, a large triangular part, with a bone in it. Use this for stock.

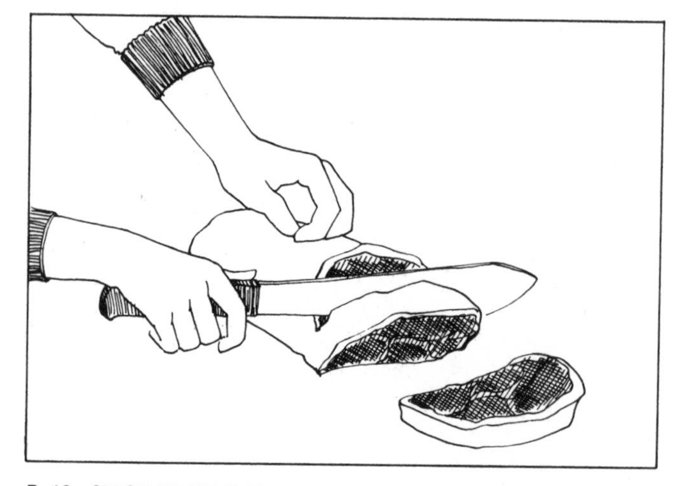

P-12. SLICING CHOPS FROM SIRLOIN

RIB AND LOIN

The rib and loin sections of the pig are left. Place the section skin side up and begin the process of removing the skin by making an incision with your knife right down the center of the entire section. Then, insert the knife blade just under the skin at the left corner and begin to glide the blade along, pulling the skin back with your other hand (P–13). The skin should come off in large pieces—sometimes even in one piece.

Now half the section is skinned (P–14). Turn the entire section around, lift up the other corner of skin, and insert knife blade. Glide and pull as you did before, keeping the knife flat so as to not cut into the meat. Remove the outer skin entirely. The skin can be used for cracklings or cut into small pieces for use in pork and beans.

Turn the section over so that the side with the skin removed is down. Take the saw and cut down, lengthwise, along the entire section approximately 1 inch below the eye to make 2 long sections (P–15). With a large knife, finish the cutting process until you reach the end of the ribs. At this point—the loin section—move blade to cut downward at a 25° angle to the very end of the bottom part of the loin section. You will then have 1 rib and 1 loin section and spareribs.

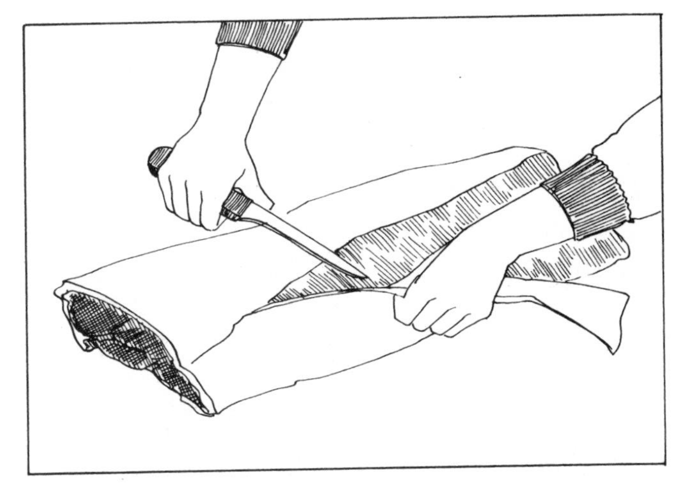

P–13. CUTTING SKIN OFF RIB AND LOIN

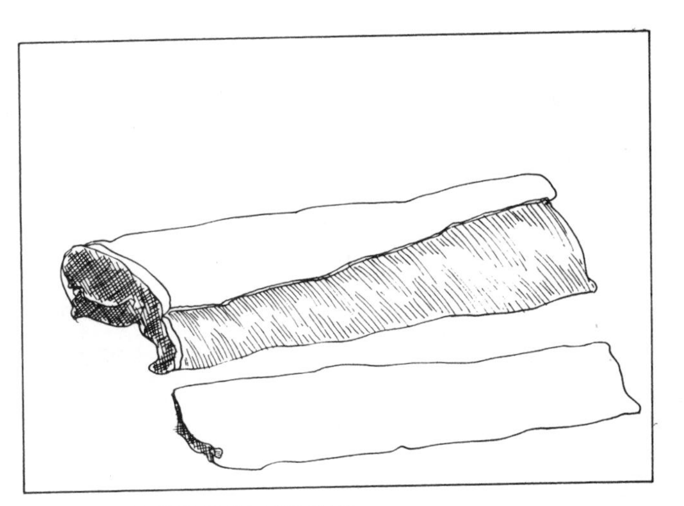

P–14. HALF SECTION, SKINNED

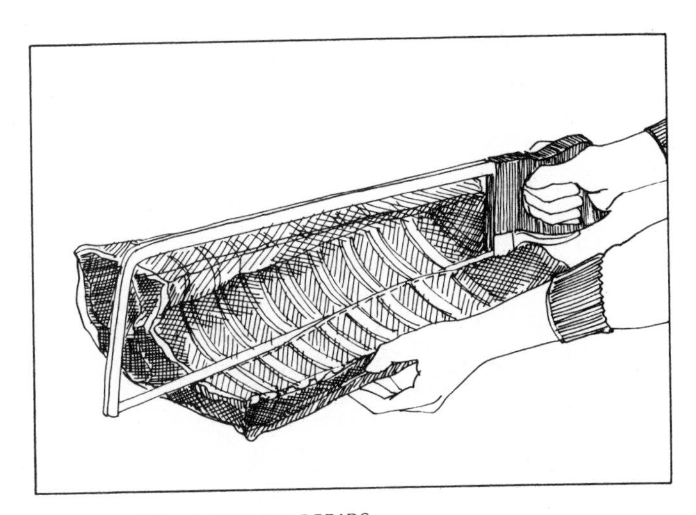

P–15. CUTTING OFF SPARERIBS

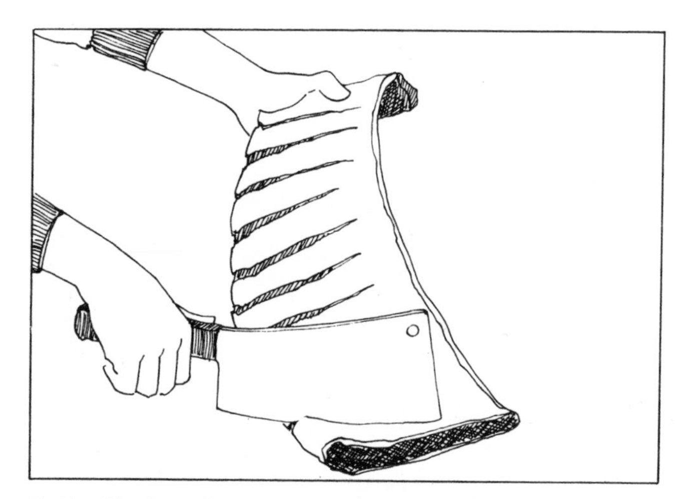

P-16. CRACKING INDIVIDUAL SPARERIBS

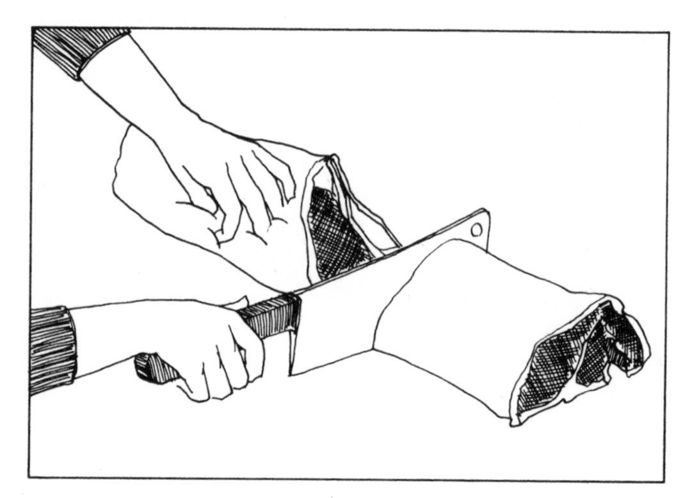

P-17. CUTTING LOIN SECTION FROM RIB AT LAST RIB

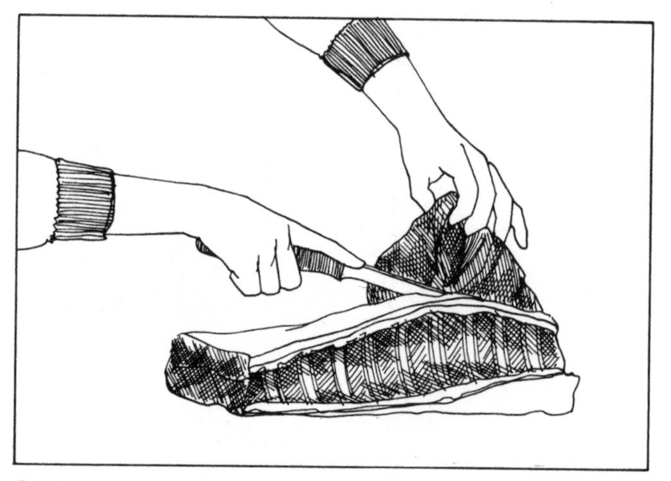

P-18. REMOVING THE BLADE BONE

SPARERIBS

Place your cleaver in between the ribs of the sparerib section. Keeping cleaver in between the ribs, chop down between each rib and the next (P–16). However, stop before you go through the bottom cartilege section, which should remain attached. These are the spareribs. Most people prefer to keep them attached so they are cooked in one unit and cut apart after cooking.

RIB ROAST AND LOIN

After removing the spareribs from one half of the pig, you have an entire loin section, which has a rib end and a loin end. Where the rib bones stop, the loin begins.

RIB ROAST

Cut the section at the end of the last rib. Begin with a knife and finish with a cleaver to chop through the backbone (P–17). The rib roast can be cracked between the bones by saw. Hold two fingers above saw blade in an inverted "V" shape, as you might hold a pool cue, but not pressed against the saw. Do not press hard with the saw; let the weight of the saw do the work for you. You must remove the blade bone at the larger end of this section. To do this, take hold of the meaty area with your fingers and give a little tug. You'll see a perforation between the meat of the ribs and the blade bone. Cut in with the knife blade (P–18), following a half-moon shape around the blade bone to free it. When it is freed, remove it completely. You now have a pork rib roast.

CENTER CUT RIB CHOPS

From the rib roast, as above, cut the section into chops, cutting between the bones. Start at the first bone, holding the bones away from you, and cut down along the line of the rib bones for center cut rib chops (P–19).

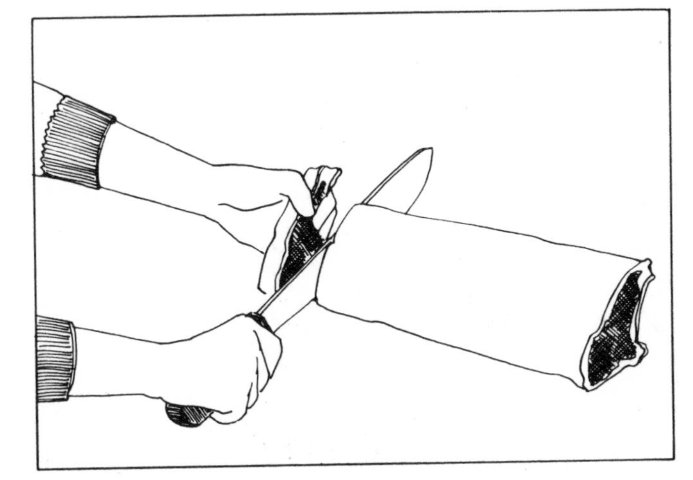

P-19. CENTER CUT RIB CHOPS

BONELESS RIB ROAST

Place the knife at top part of the rib bone at the end. Run the blade along the top of the ribs, pushing back the fat and the meat as you do. Keep running the knife along the ribs, cutting toward the backbone area over the ribs. Completely free the meat from the ribs (P–20). As you reach the chine bone area, work the blade up and around the chine bone to free the meat (P–21). When you've freed the chine bone and the ribs, lift them away. What is left is a boneless rib roast. Roll and tie the meat neatly if necessary.

Crack the rib bones that remain or leave them as they are. You can use them either way as a rack for the boneless roast (P–22). Boneless roasts from very small young pigs do not need to be tied. Simply place on a rack and roast.

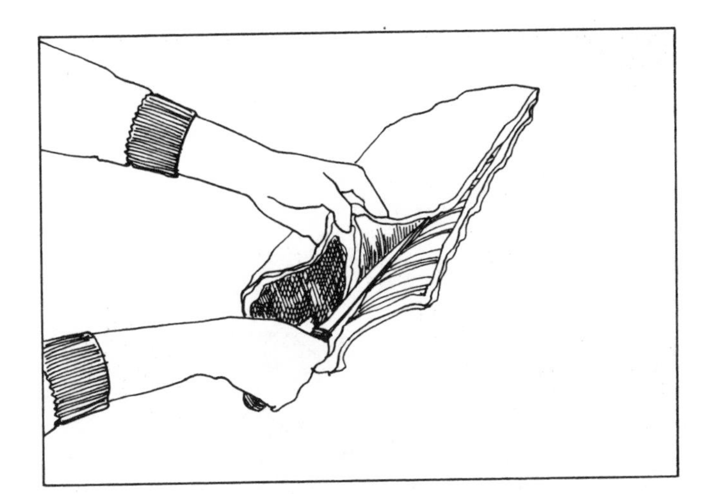

P-20. BEGINNING TO BONE

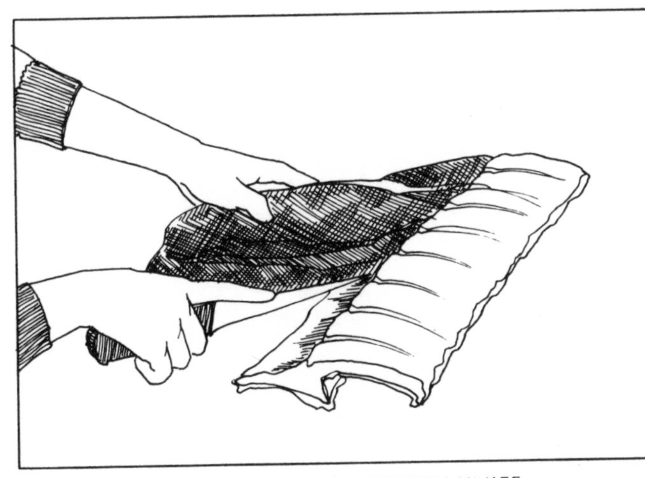

P-21. FOLLOWING CHINE BONE WITH KNIFE

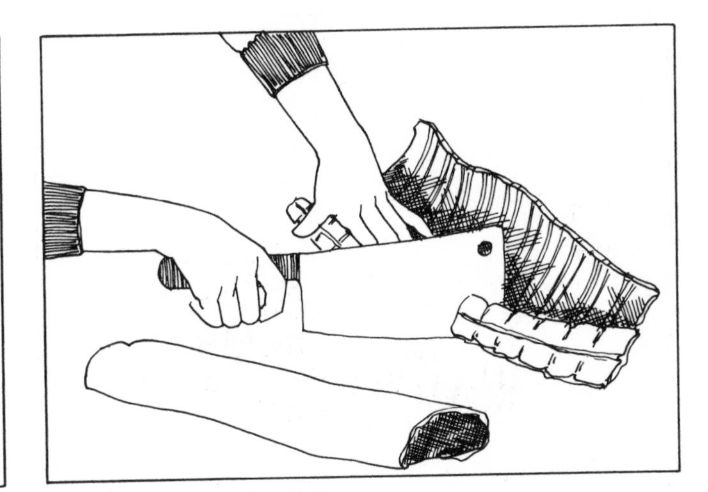

P-22. CRACKING RIB BONES

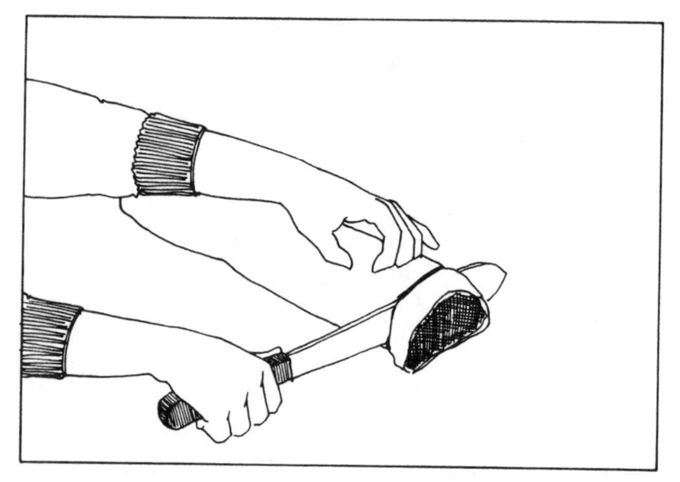

P-23. CUTTING BONELESS CHOPS

P-24. ROAST LOIN OF PORK WITH BONE

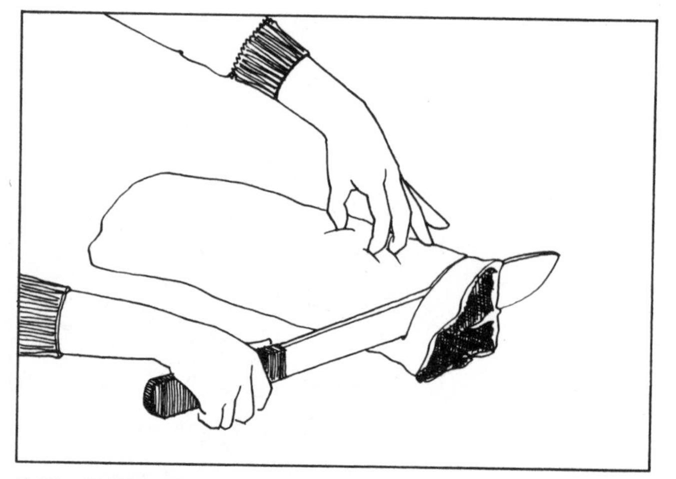

P-25. CUTTING LOIN CHOPS

BONELESS CHOPS

Cut chops from the boned roast according to the thickness desired (P–23).

PORK SCALOPPINE

Cut thin pieces from the boneless rib or loin section. Pound and flatten in the usual manner for scaloppine.

LOIN ROAST

This is the loin section left after separating the rib half of the entire loin section. It is sometimes called the T-bone section. Simply crack or partially saw the top backbone section of the loin (for easier cutting after it is roasted) and the loin is ready for the oven (P–24).

LOIN CHOPS

Cut chops from the loin section according to the thickness you wish (P–25).

DOUBLE-CUT PORK LOIN CHOPS

Cut chops from the loin section in double widths for added thickness.

STUFFED PORK LOIN CHOPS

Use thick chops for this. Insert knife blade and make 1-inch slit toward bottom of the chop (P–26). Push the knife through the chop until you hit the T-bone. Then run the knife upward very slowly *inside the chop only,* moving the blade upward another 2 inches. Then pull the blade out. Place your finger in the incision, opening the interior of incision further but keeping outer incision narrow (P–27). Stuff the incision after it has been opened inside. No need to skewer it shut.

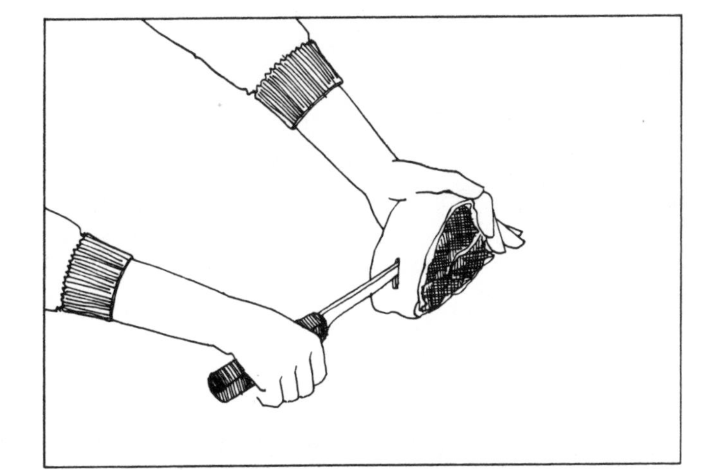

P-26. MAKING POCKET FOR STUFFING

CROWN ROAST

Follow the process exactly as detailed for crown roast of lamb (page 27).

PORK KABOBS

Cut cubes for kabobs from loin section, using the long fillet section if possible. This lies just under the backbone and is shaped like a long, narrow triangle resting on its side with its base the end of the loin section.

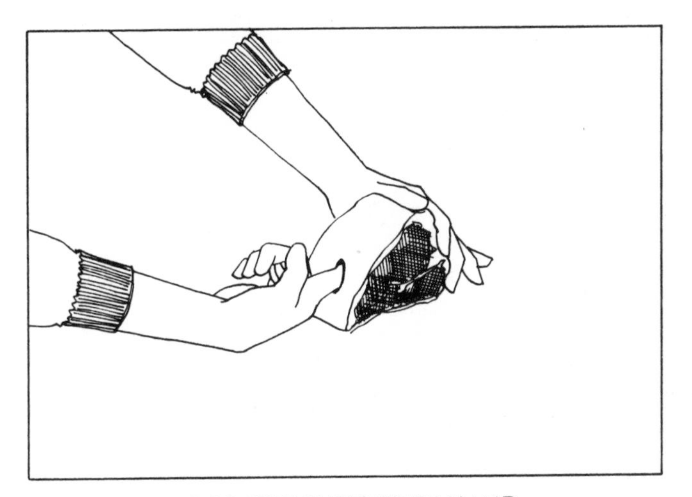

P-27. OPENING POCKET FURTHER BY HAND

BUTTERFLIED PORK CHOPS

Cut these from the boneless rib section. Cut a chop from the boneless meat, then cut down the center of the chop, *almost but not quite* cutting it in two (P–28). Spread the 2 sections of the chop out flat.

PORKBURGERS

Use all the scraps and pieces of meat left over for grinding.

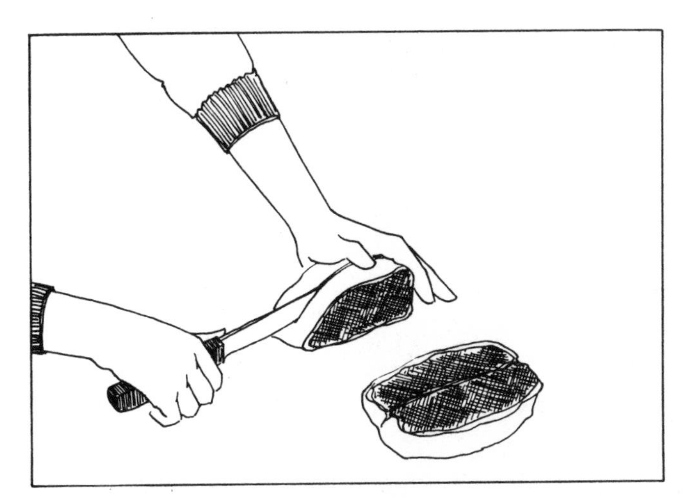

P-28. BUTTERFLYING PORK CHOPS

MISCELLANEOUS

PIG'S FEET

Pig's feet must be rinsed with cold water and simmered slowly for eating, usually about 2 hours, or until tender. In the market they are often sold pickled in vinegar, a process you can do if you have a knowledge of pickling.

CHITTERLINGS

These are the heart, kidneys, liver, and sweetbreads sliced or chopped and usually fried. They are rich in taste and nutritional content. These may often be purchased separately from the whole pig.

PIG'S LIVER

Remove liver when you first bring home your pork and set it aside. It may be sautéed and/or used as the base for any variety of pâtés.

BACON

Bacon is taken from the same section of the pig as spareribs. Because of the need for smoking and curing, we do not suggest that you attempt to cut and cure bacon in your home. Bacon, both the slices and the whole slab variety, can be purchased in the marketplace, with or without chemical additives and with less work and cost than it would take to process it at home.

7

Veal

Veal, the fatted calf of Biblical times, has become popular again in this modern age. The ancient Jewish and Egyptian cultures knew the pleasures of veal, but a combination of social, political, and religious factors worked against veal's becoming a popular dish in the West for many centuries. The young calf was considered primarily a dish for special occasions, and thus not served at ordinary mealtimes.

But, in the worldly and sophisticated climate of Renaissance Italy, particularly Tuscany, veal began to come into its own again. The wealthy Tuscan merchant Francesco de Marco Datini made veal an important part of all his lavish banquets—his doctor having told him: "Place it in your belly, in every way you can, for you can have no more wholesome victuals."

That advice is equally sound today. Veal has very little fat content and has been praised as having the subtlest flavor of all meats. In the United States veal is considered veal until the calf is six months old, at which point it is defined as beef. In purchasing veal in bulk, learn to look at the whole veal even if you will be buying it in quarters. Different wholesalers favor different strains of cattle and you will, in time, develop your own preferences. We particularly like the Aberdeen-Angus and the Hereford in both beef and veal.

Note the conformation of the veal. The back should be broad, the rib cage rather barrel-shaped. Shoulders should be short and chunky, the neck short and thick—a compact animal with a look of youth. The meat should be as white as possible, the fat even whiter. Dark fat means the veal is older than it should be and consequently not as tender as it should be. The bones of good veal should be red. If they are brown, don't buy it.

We do not recommend purchasing veal in any way except in 4 or at the most 2 sections. Practical handling considerations as well as storage (and freezing) dictate this. Below are the major sections of veal. Cut the individual pieces you want from the following sections:

FOREQUARTER

SHOULDER AND FORESHANK

Boneless Shoulder Roast, also called Rolled Veal Shoulder
Blade or Arm Steak, also called Shoulder Steak
Shoulder Scaloppine
Stew
Ground Veal
Osso Buco

BREAST

Roast Breast of Veal
Stuffed Breast of Veal
Ground Veal

RIB

Rack of Veal Roast
Rib Chops
Rib Cutlets
Petite Crown Roast

NECK

Roast Neck
Stuffed Neck
Stew

HINDSADDLE

LOIN

Loin Veal Chops
Stuffed Double Veal Chops
Single Veal Kidney Chops
Boneless Loin Roast
Veal Loin Roast
Boneless Loin Veal Chops

LEG AND RUMP

Pin Bone or Hipbone Veal Chop
Veal Sirloin Roast (also called Veal Leg Sirloin Roast)
Veal Bottom Round (also called Veal Round Roast)
Scaloppine
Stuffed Roast Boneless Veal

EYE ROUND

Medaillons de Veau
Roast Eye of Veal

SHORT SIRLOIN/FLANK AND TOP SIRLOIN

Veal Kabobs
Scaloppine
Veal Fondue
Roast Sirloin

TOP ROUND

Veal Steak
Veal Cutlets
Scaloppine

SHANK

Osso Buco

SET-ASIDE PIECES

Stroganoff
Veal Stew Elegante
Vealburgers

Even when buying meat in bulk, you may want to buy smaller bulk sections, dictated by your own needs and space limitations. You may want to buy separately only the shoulder and foreshank of the forequarter, or only the breast or rib or neck section. You may wish to purchase only the loin from the hindquarters or perhaps only the rump section, which will have the top round and bottom round. Most of the major markets selling bulk or wholesale sell these sections separately or will cut them for you.

If you have purchased only one or two of these smaller sections, simply use the methods described here for cutting and preparing them. However, in the event that you have the space and the inclination to begin at the beginning, we use the basic forequarter as our starting point. Also, the entire forequarter will generally be the most economical purchase.

FOREQUARTER

The forequarter is one half of the front half of the veal (V–1). The upper foreleg is on this section. This is the section which you will cut into 4 primal cuts. The forequarter also happens to be the section used in kosher butcher operations.

The first step is to place the back of the forequarter next to you, with the leg pointing away from you. Using your hand, feel along the top of the section until you find the top of the shoulder bone (the blade bone). With your large knife, make a vertical incision right next to the shoulder bone, going down only as far as the membrane (V–2), the very broad whitish area *beneath* the first layer of meat. At the top of the shoulder bone you will see a piece of white cartilege, thicker than the membrane. Cut through this to loosen it.

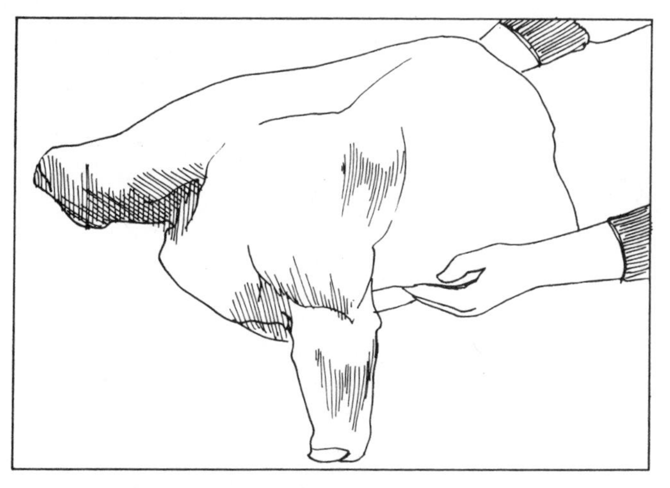

V-1. THE FOREQUARTER OF THE VEAL

V-2. WITH LEG AWAY FROM YOU, CUTTING DOWN TO MEMBRANE NEXT TO SHOULDER BONE

Make a cut at right angles to remove the skin over the top of the shoulder (V–3), cutting only deep enough to remove the skin. When you reach the forefront of the shoulder bone, continue cutting the flap of skin until it comes loose between the neck and the shoulder. When the skin is loose, the shoulder will move easily, and will be ready to come off (V–4).

Turn the piece around so the leg faces you. Lift slightly, hold leg up straight, and continue cutting *underneath* leg. Keep the knife close to the line of the shoulder, cutting between the breast and ribs and the shoulder. You will be able to cut through almost all of the membrane. You should now be able to pull the leg all the way over and see the flat underside of the shoulder bone. Continue to cut away the membrane and skin until the shoulder comes loose. The shoulder and foreshank section can now be set aside.

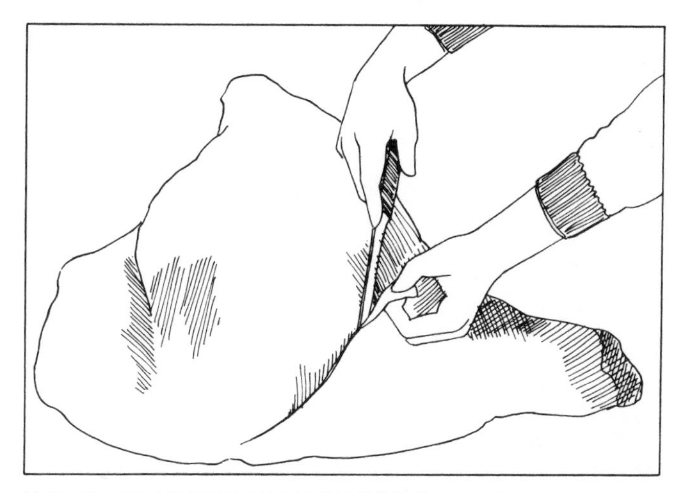

V-3. CUTTING SKIN AWAY ON TOP OF SHOULDER

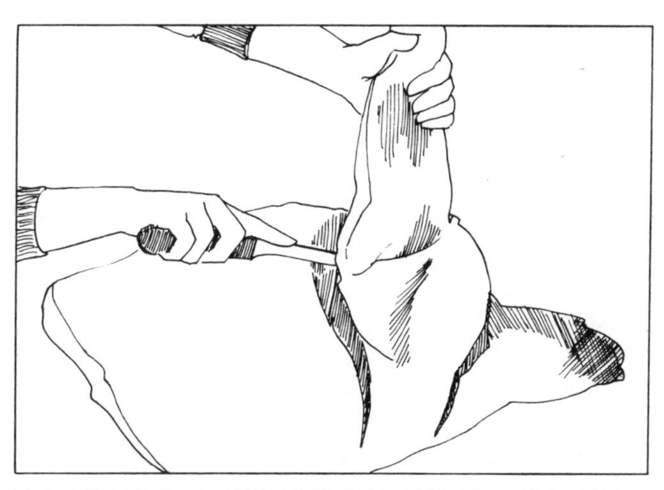

V-4. LOOSENING SHOULDER BY CUTTING UNDER SKIN

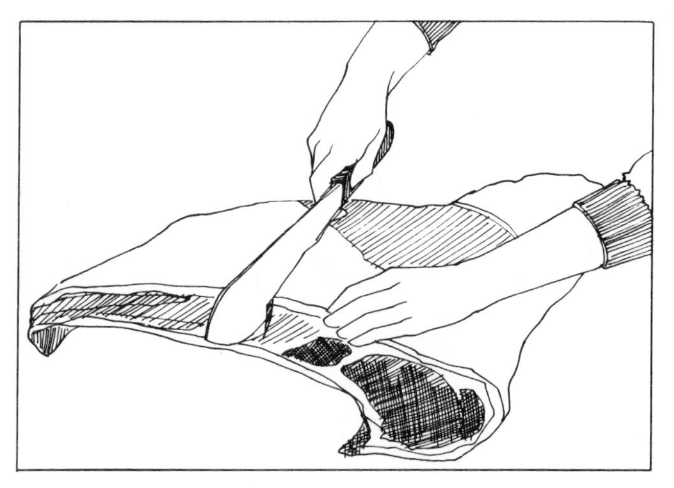

V-5. MARKING TO REMOVE BREAST FROM NECK AND RACK

Next, remove the breast from the neck and chop (rack) section. Place the underside down, which will leave the bone side down, and measure off the breast section. Looking at the breast end of the piece, you will see the curved bones of the rib and the meat (eye) of the rib chops. Measuring up from the end of the pink meat 3 inches, make a small incision as a mark for yourself (V–5). Then go to the other end of the piece and feel with your hand until you find the base of the neck itself (V–6). Make another incision mark at this point.

Use the knife to make a cut down the line from mark to mark, or end to end. Then use the saw to cut through the bones deeper inside this line (V–7). Let the saw do the work—do not use too much pressure as you hold onto the rib bones with your other hand while sawing. When finished, you will have 2 pieces. One is the breast, with the ribs; the other, the neck and rib-chop section, also called the rack.

Next remove the neck. From the neck, count down 3 bones. Place your knife between the 3rd and 4th bone and make an incision all the way down until you strike bone. Then use the cleaver to chop off the entire piece (V–8). This will separate the neck and the rib chop (short rack) sections. These are the 4 primal cuts: the shoulder and foreshank; the breast with the ribs; the rack; and the neck.

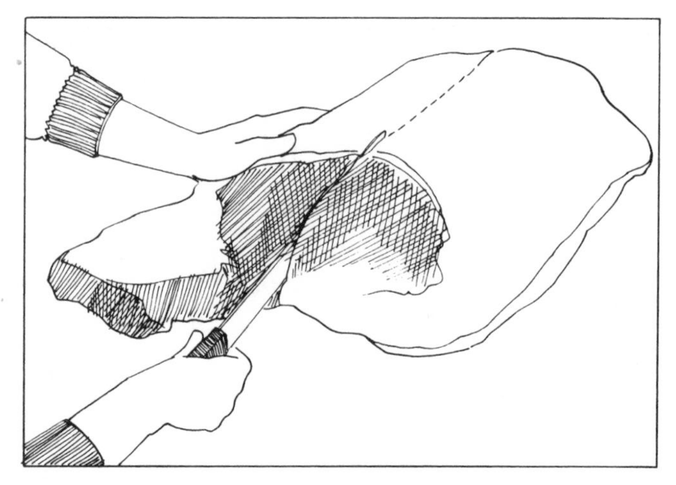

V-6. MAKING INCISION AT OTHER END OF PIECE

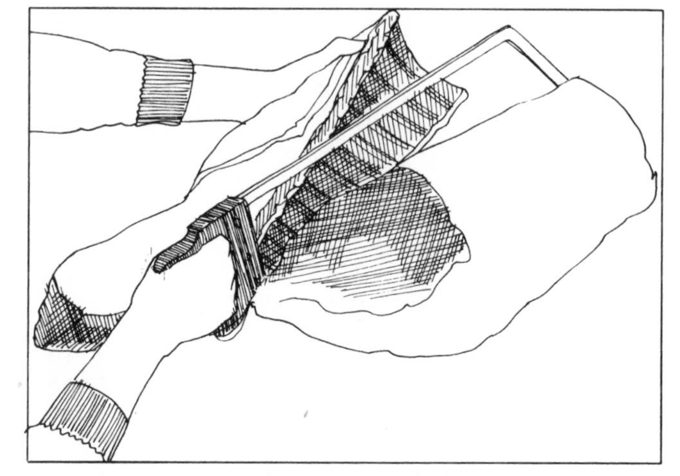

V-7. USING SAW TO CUT THROUGH BONES

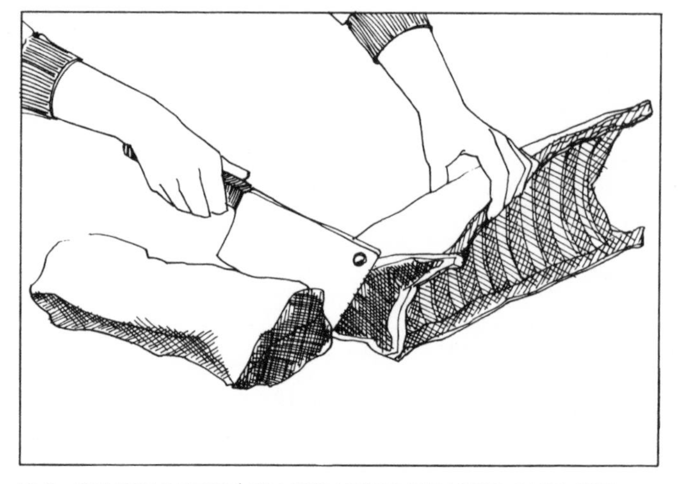

V-8. DISCONNECTING NECK BETWEEN 3RD AND 4TH BONES

SHOULDER AND FORESHANK

Cut the foreshank from the shoulder by moving the foreleg (shank) until you feel the knee joint. Cut through the knee joint with the knife as far as you can, finishing off with the saw (V–9). The shank piece that is cut off will be at the end of a piece of shiny white leg bone, the ball-and-socket joint.

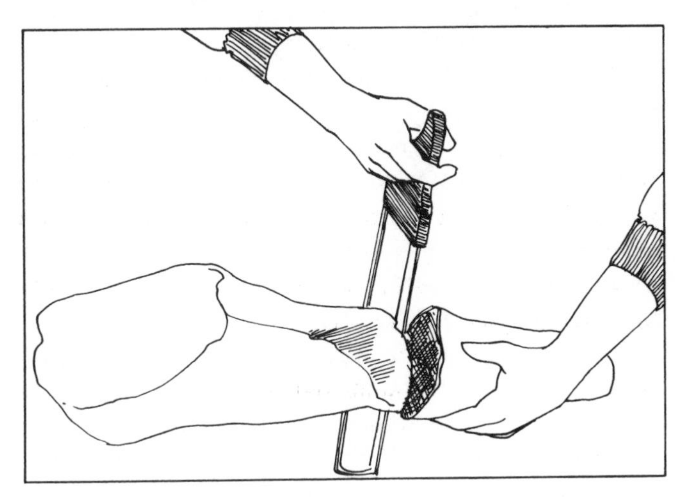

V-9. CUTTING FORESHANK AT KNEE JOINT

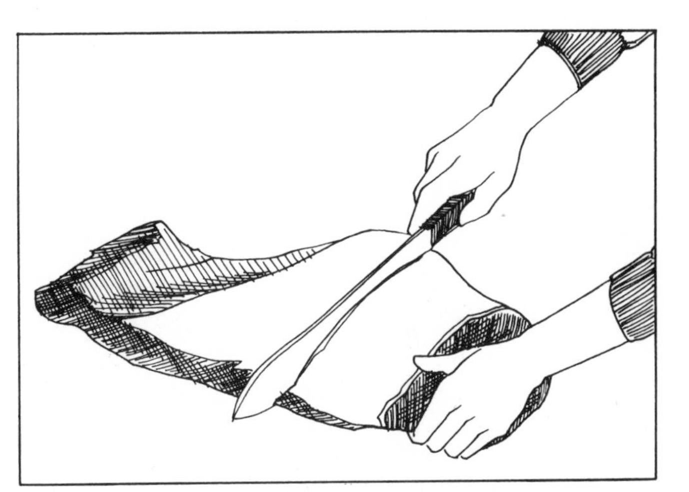

V-10. CUTTING SHOULDER FOR STEAKS

BONELESS SHOULDER ROAST

Place the shoulder piece with the round, shiny end of the leg bone facing you. Work the knife around the leg bone, in between the leg bone and the kneecap (V–11). Continue cutting until you have freed the entire front half of the leg bone. With the front part exposed, turn the meat to whatever angle best suits you until you can wiggle the leg bone freely. Continue to cut around the large, smooth bulbous end of the leg bone and remove it completely (V–12). Remove the kneecap by cutting around it (V–13). *Note:* In the cutting, you will invariably cut off small scraps of meat. Set these aside and save.

Next, remove the blade part of the shoulder bone. You will clearly see the end of the blade bone. First start to remove the meat from around the bone. Keep the knife flat along the top of the bone (V–14) as you work to lift the meat and expose the entire top part of the bone. When you have exposed the top of the blade bone, slip the tip of the knife flat against the *underside* of the blade bone (V–15) and

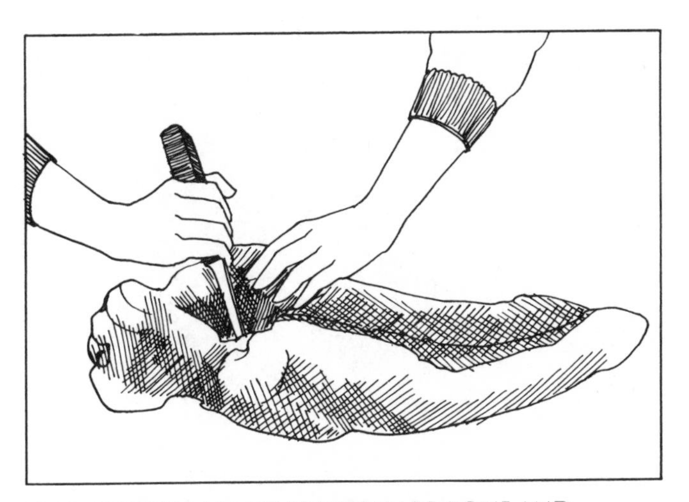

V-11. WORKING KNIFE BETWEEN LEG BONE AND KNEECAP

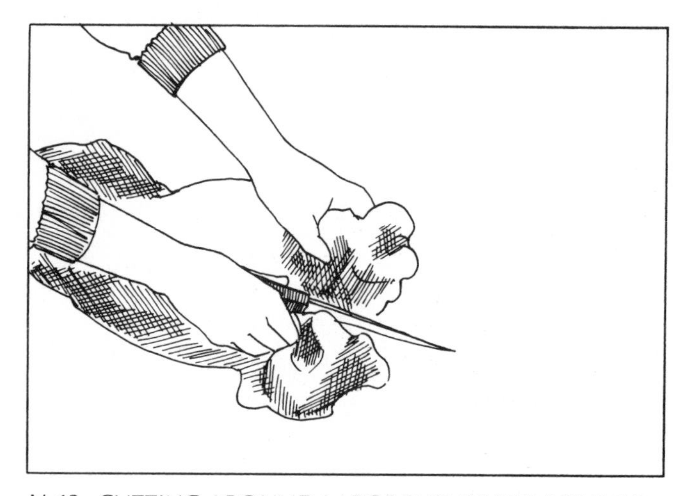

V-12. CUTTING AROUND LARGE END OF LEG BONE TO REMOVE KNEECAP

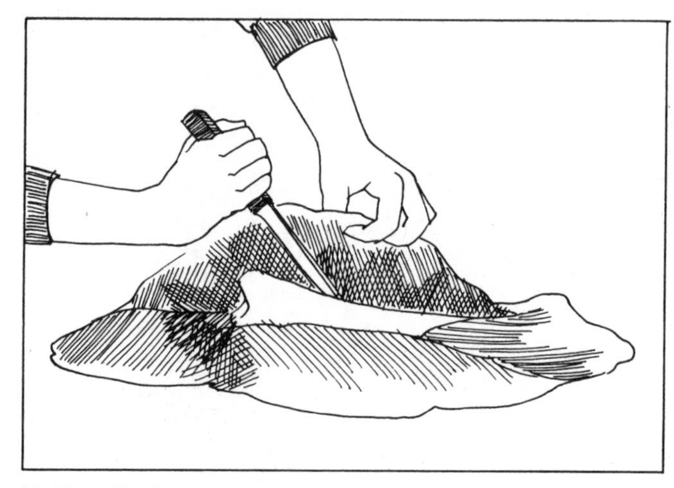

V-13. USING KNIFE TO BONE OUT UPPER ARM BONE

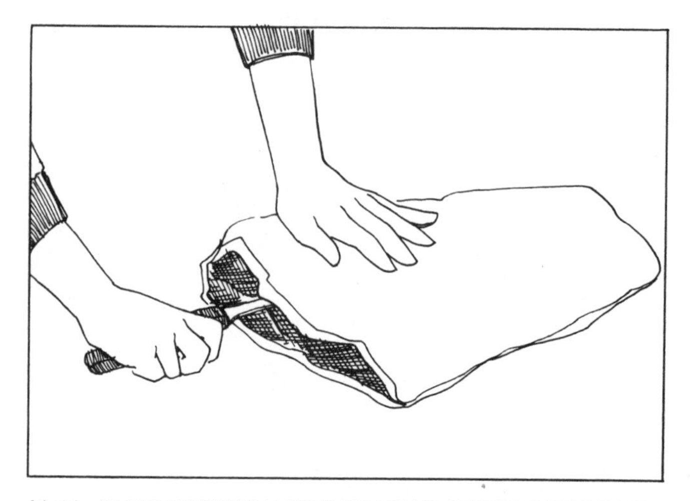

V-14. KNIFE POSITION FOR BONING OUT BLADE BONE

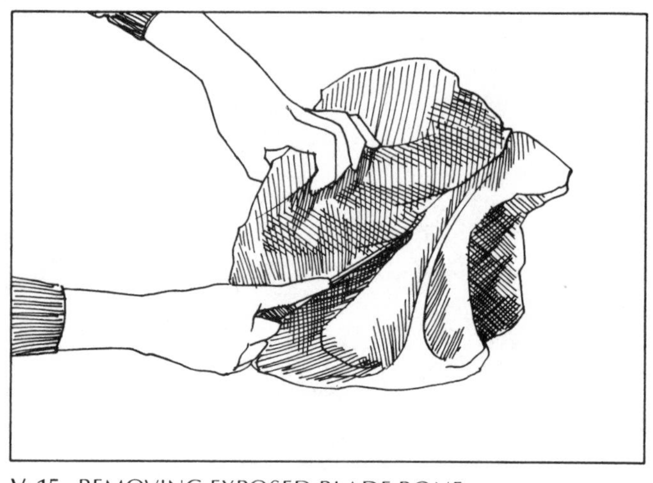

V-15. REMOVING EXPOSED BLADE BONE

cut along the *underside* of the blade bone at the edge. You will only be able to go in a short way before you hit a shelf of bone. You must go *under* this shelf, letting the knife follow the contours of this part of the bone until you have cut away the oblong, conelike section of meat along the bone. Set this aside. The whole side of the bone will now be visible.

Turn the entire piece over. Run the knife along what is now the top of that same bone. Make an incision alongside the bone, following the line of the bone, which will now be sticking up toward you. Clear all the meat off the flat of the bone and keep cutting along the edge of the flat shoulder-blade bone until you have cut out the entire bone.

What remains is a large, floppy piece of meat. Our own skinning technique, which we have found best over the years, is to slip a knife between the skin and the meat and run the blade with the facing edge slightly upward to remove the skin and outer membranes. When this is done, the piece should be given a general cleaning, removing all excess fat, turning the piece over until it is trimmed and cleaned on both sides.

Press the meat together. To roll, start by tying the meat in the center. Then make two more ties, one at each end (V-16). Work the others toward both ends from the center, making ties 2 inches apart first, then filling in until you have ties 1 inch apart. As you tie, your meat will assume a rolled shape. You must tie a shoulder roast more thoroughly than most other rolled roasts because the consistency of the meat requires it.

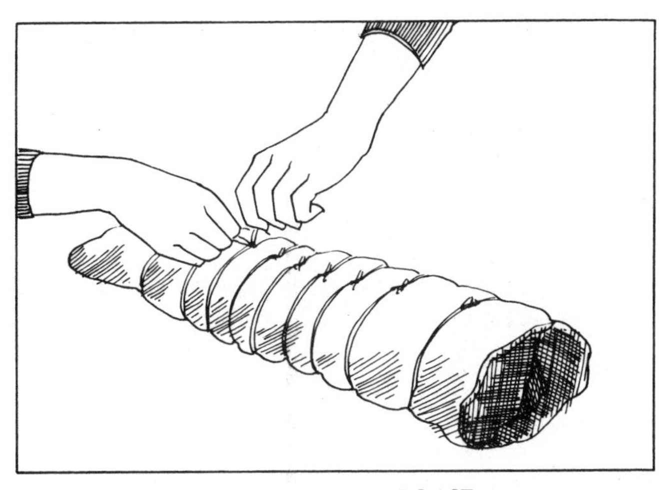

V-16. TYING SHOULDER OF VEAL ROAST

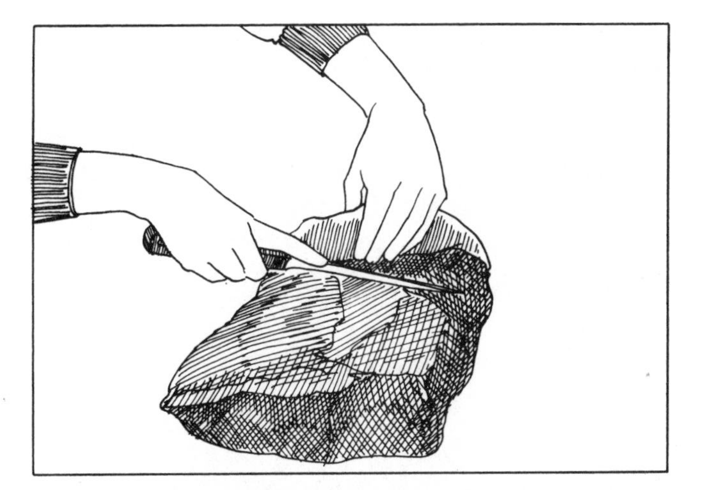

V-18A. SECTIONING OUT LARGE PIECES

BLADE (ARM) STEAKS

These are cut from the shoulder. Simply cut them to the desired thickness. Find the blade bone at the front end of the shoulder section (a flat, medium-sized bone), and cut steaks from this end, first using the knife, and finishing with the saw (V–10).

SHOULDER SCALOPPINE

Scaloppine can be cut from the boned shoulder but not without taking extra care. Scan the meat and you will see seams between the very thick parts (V–17). Cut along these seams, which look like a thin film, gliding your knife along the thick veins and cartilege to remove them (V–18A,B,C). Use the edge of

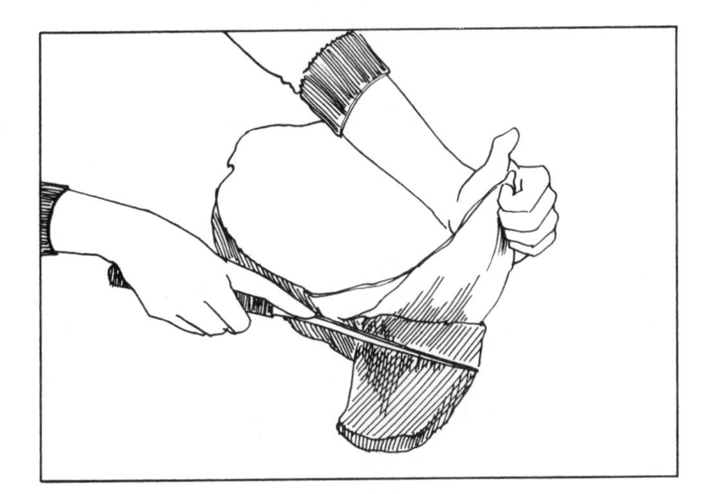

V-18B. SKINNING AND TRIMMING

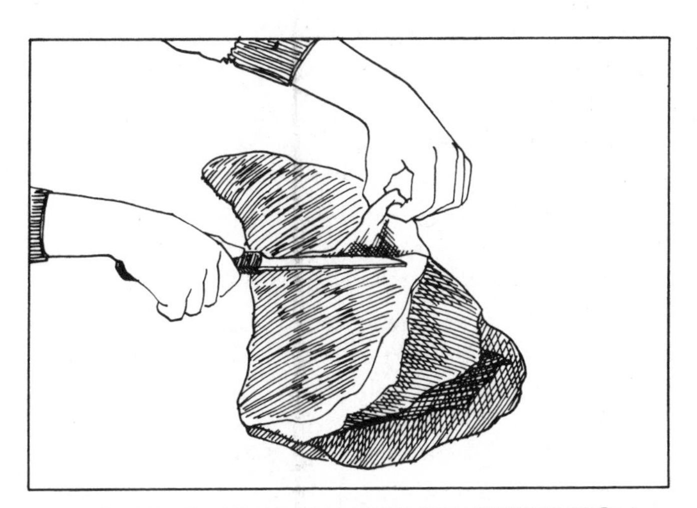

V-17. CUTTING ALONG SEAM FOR SCALOPPINE FROM SHOULDER

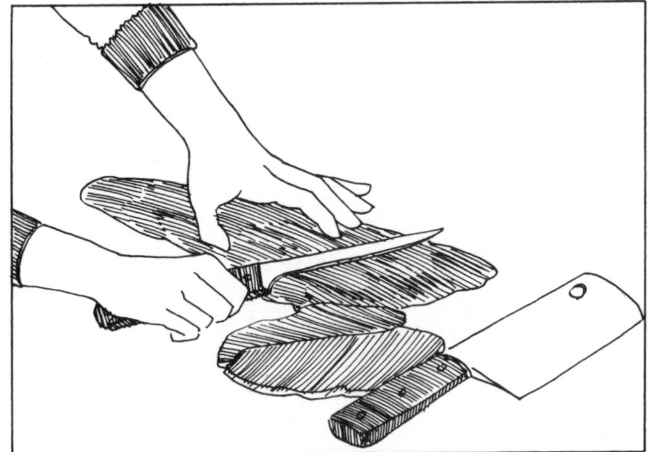

V-18C. SLICING FOR SCALOPPINE; POUND SLICES FLAT WITH CLEAVER

the blade in small strokes rather than one long cut, removing the fat and membranes until you have a piece of meat trimmed of these gristly elements. Shoulder scaloppine can be cut only from this trimmed piece, and should be cut in thin slices (scallops) *against the grain.* (See the section on hindsaddle for how to cut and prepare scaloppine.) Save the pieces of meat you have trimmed away and add to the other scraps you have put aside.

OSSO BUCO

Although foreshank Osso Buco is not our preference, it can be used, and some people do prefer it. At one end of the foreshank the top of the bone of the joint will be exposed. Go in about 1 inch and with the saw, cut off that end (V–19). Then make 3 cuts, 2 inches apart, in the rest of the foreshank to get the Osso Buco pieces. A large end of the hock will be left (V–20). This can be used for stock, gelatin, or soup thickening. The smaller end, which was cut off first, may be used for the same purposes.

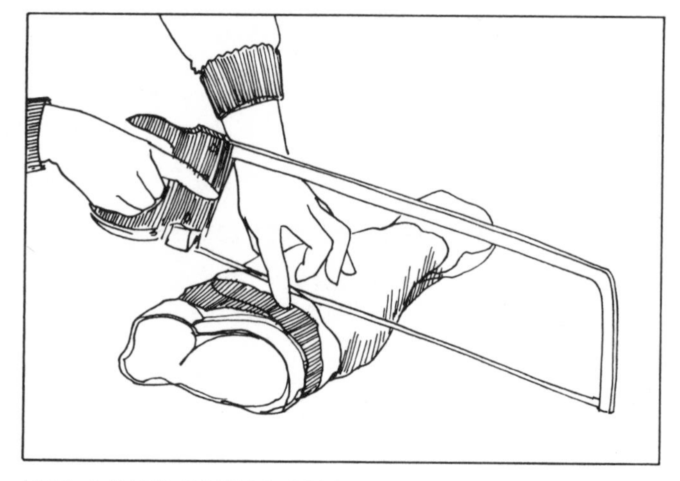

V-19. 1-INCH CUT FOR OSSO BUCO

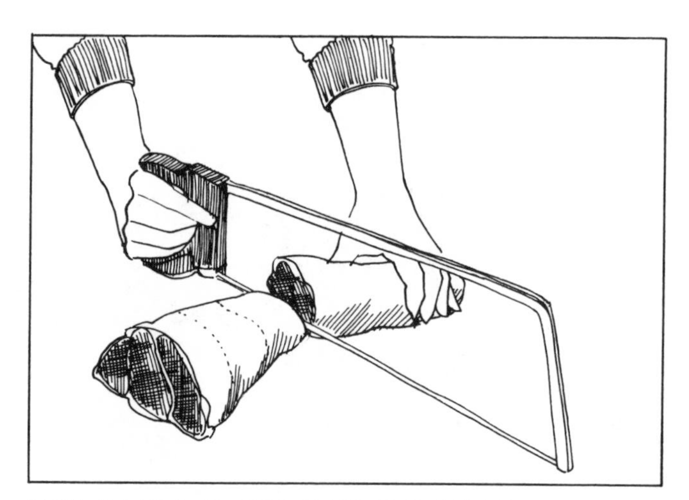

V-20. REMOVING OSSO BUCO FROM HOCK

BREAST

The breast is the fattiest portion of the veal. It may be cut in half for easier handling or convenience or storage (V–21). There are approximately 12 ribs per breast. Cut down in the center between the bones to make 2 rib sections of 6 ribs each.

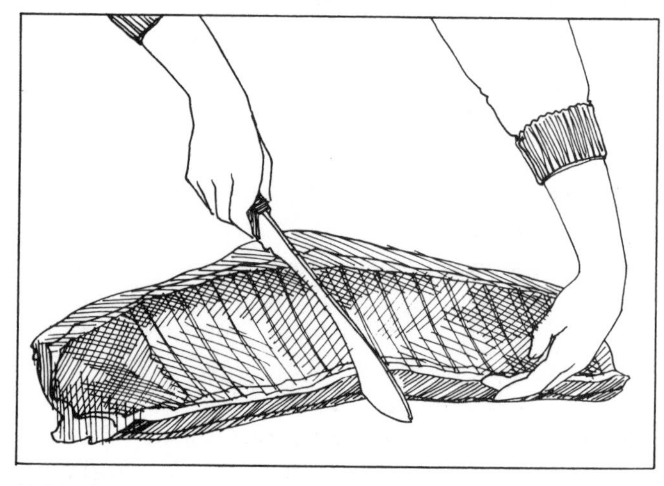

V-21. CUTTING BREAST IN HALF

ROAST BREAST OF VEAL

Place bones down and remove all heavy excess fat. Trim on the top and the sides, but leave a slight, thin outer covering of fat. Turn the piece over and do the same on the underside.

Position the saw between the rib bones and against the dark, oblong section of the backbone and saw down to crack the backbone. Saw through the bone and the cartilege, making sure that each cut you make is parallel to and between the rib bones (V–22). You will, when finished, have made 11 cuts for the whole 12-rib section, 5 cuts for the 6-rib half-section. The breast of veal is now ready for roasting. This section can also be cut for spareribs (V–23).

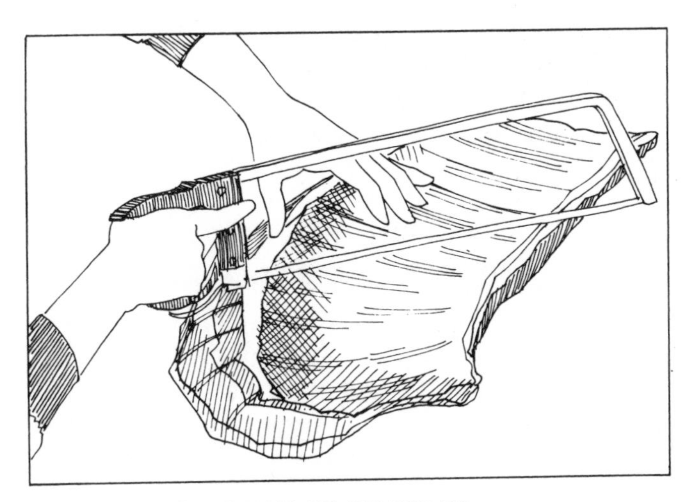

V–22. SAWING PARALLEL TO RIB BONES

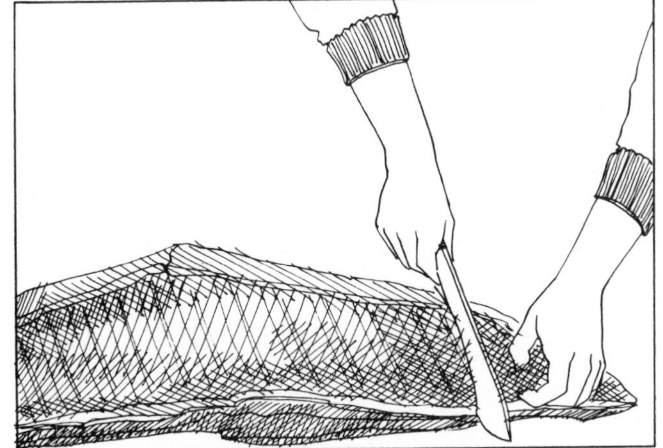

V–23. CUTTING THROUGH FOR SPARERIBS

STUFFED BREAST OF VEAL

Place bones down with the front part of the breast facing you. Insert the 6-inch knife flat between the bones and the meat areas, pressing against the bone in the center. Keeping the knife blade flat, follow the contour of the breast to start to open a pocket in the meaty section (V–24). Work the blade back and forth as you move forward approximately 3 inches. Work the 6-inch blade in to the hilt, withdraw and insert the 12-inch blade into the pocket opening you have made. One edge of the breast will be thicker than the other. Work the knife inside the pocket along that thicker end, keeping the blade flat with the edge toward the thin part of the breast. Work carefully to widen the pocket until the knife almost disappears. As you withdraw the knife, keep the blade edge toward the thin section, and widen the pocket as you withdraw the blade.

Turn the meat over and crack the backbone as you did for plain roast breast of veal, but with this important difference—you must carefully sever the cartilege with the knife, putting aside the saw after your initial cut so that you will not cut into the pocket. Stuff the pocket and sew or skewer closed (V–25). *Note:* We do not suggest using breast meat for stew because it is too fatty.

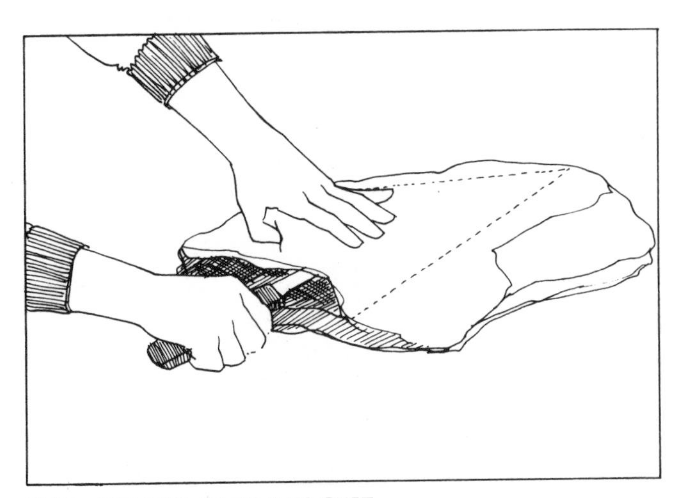

V–24. CONTOURS FOR POCKET

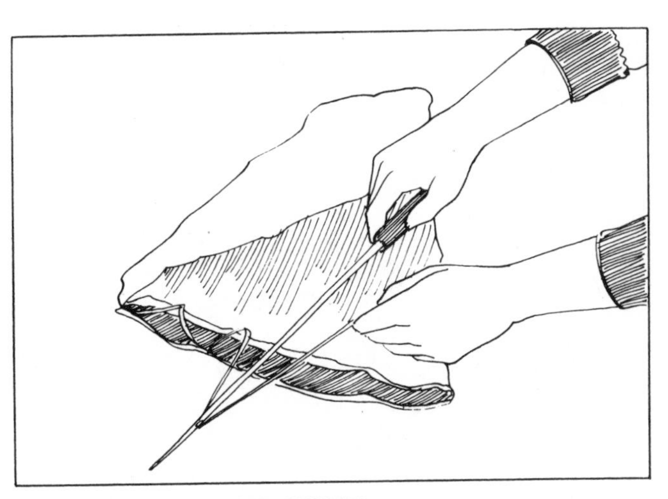

V–25. SEWING STUFFED BREAST

GROUND VEAL

Grind all the scraps of veal you have gathered in your cutting and use in any way you would use ground beef.

RIB

RACK OF VEAL ROAST

Take the short rack section which you separated from the neck and place the backbone down so that the ribs curve up away from you. Run the knife down the top edge of the ribs, going along the rib bones to cut off the membranes and fat on top of the meat (V–26). Now cut down until you can lay back the entire section. To do this, turn the piece over so that for this procedure the backbone is up. Using your saw, cut horizontally along the edge of the backbone to remove the chine bone (V–27), then use the knife to remove all jagged and split pieces of remaining backbone. These are connected by a strong milky-yellow tendonous material. Cut through this and remove it (V–28). You now have a rack of veal roast. Turn meat side down and crack between the ribs very lightly. The rack is ready for the oven (V–29). Roast bone side down.

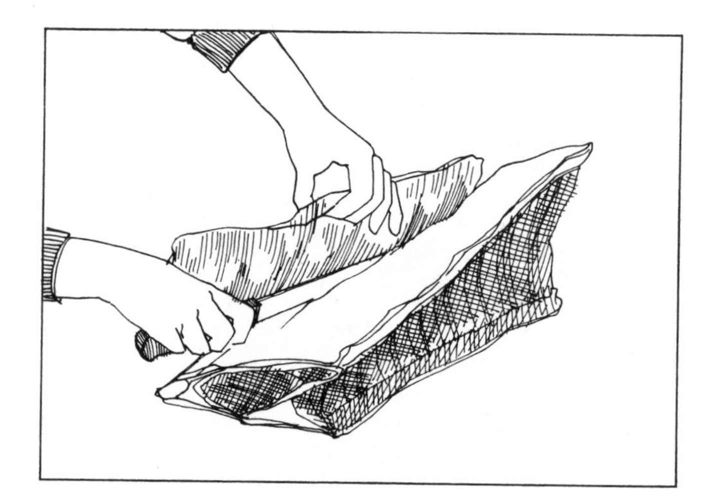

V–26. REMOVING MEMBRANES AND FAT

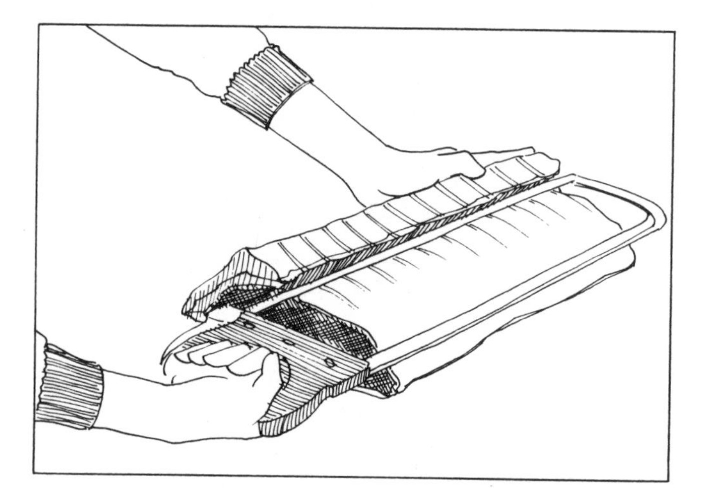

V–27. POSITION OF SAW FOR REMOVING CHINE BONE

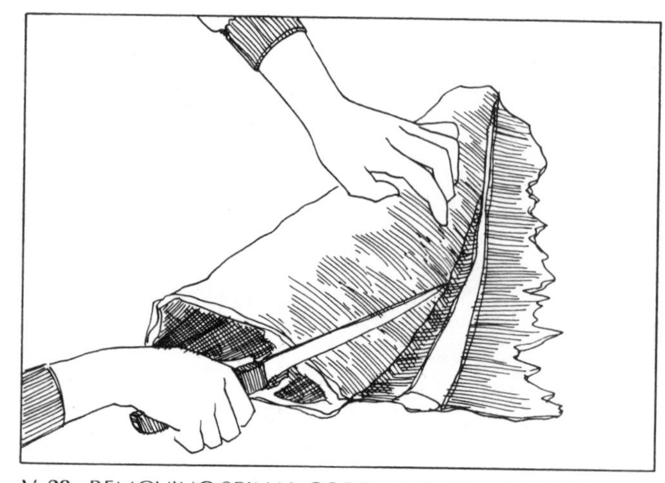

V–28. REMOVING SPINAL CORD AND REMAINING BONE FRAGMENTS

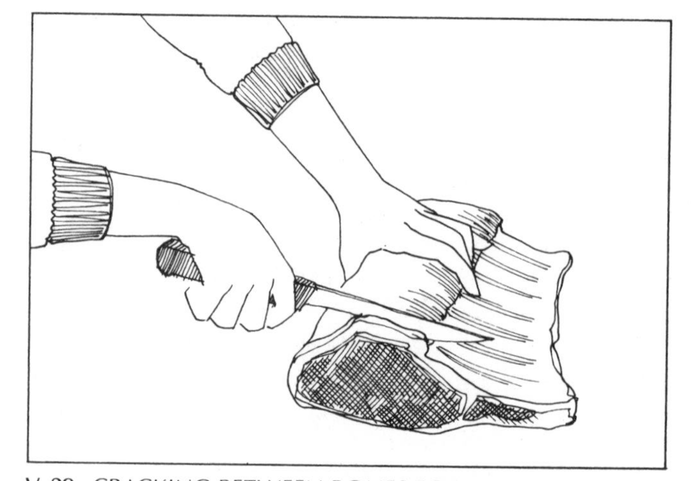

V–29. CRACKING BETWEEN BONES FOR ROAST

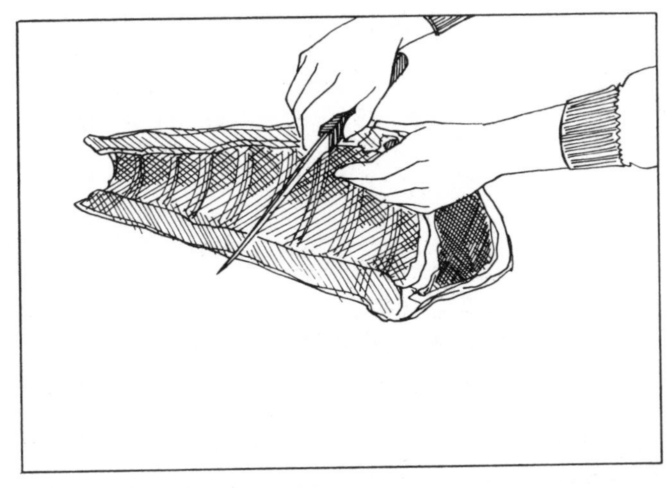

V-30. CUTTING RIB CHOPS

RIB CHOPS

Cut from rack according to desired thickness (V–30).

RIB CUTLETS

These are chops that are cut thin, ¾-inch thick at most. When cut, flatten them further between sheets of wax paper, but leave the bone unflattened.

PETITE CROWN ROAST

Follow the procedure for crown roast described in the Lamb section. However, with the veal crown, you must crack the bones deeper than you did for the veal rack roast—about 1½ inches deep. You will have to even off the tops of your bones by cutting them evenly, which can be a somewhat tricky process. For tying the petite crown roast, follow the instructions given on pages 000 for lamb crown roast (V–31 and V–32).

V-31. FRENCHING BONES FOR PETITE CROWN ROAST

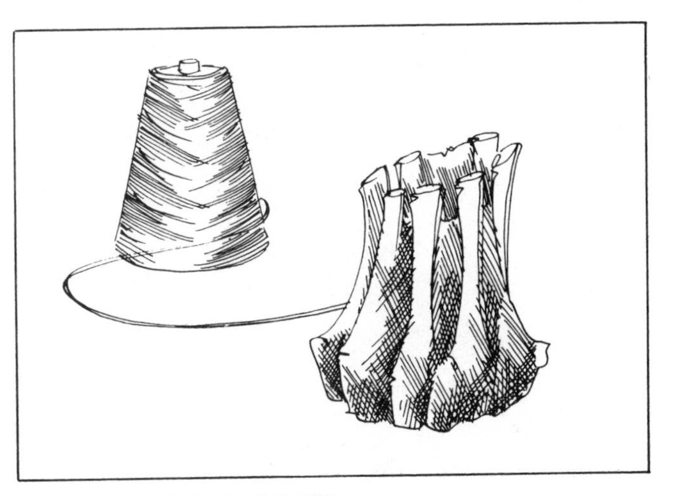

V-32. PETITE CROWN ROAST

NECK

The neck is not as soft in texture as the rest of the veal but it has less fat than the breast and we feel it makes a nicer roast. It requires more cooking time and is best done covered for half the time to make it tender, and uncovered for the other half to give it a delicious, roasted flavor.

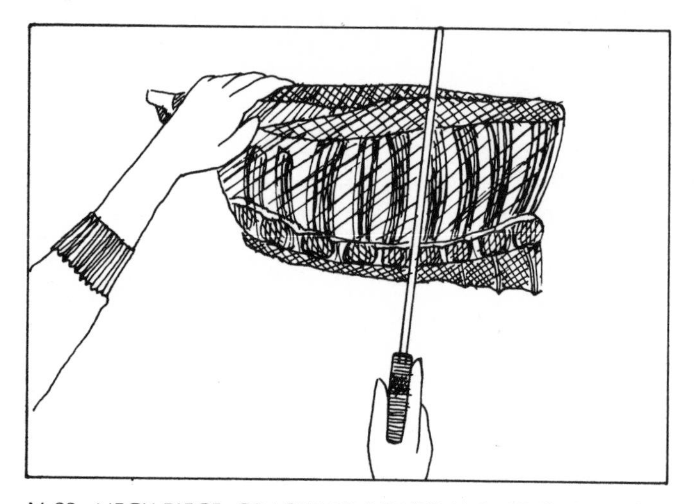

V-33. NECK PIECE: CRACKING BONES ALONG CARTILEGE

ROAST NECK

Place the neck bone side up. Use the saw to crack the bones along the white cartilege between the bones (V–33). When finished, you are ready to roast it.

ROAST STUFFED NECK

To make a pocket for stuffing, turn the neck bone side down. Proceed carefully. You'll see that the neck piece has two ends, one bony and one meaty. Place the knife in the center of the meaty end and cut right through to the bones at the other end, withdrawing the knife as you hold the blade at an angle. Reverse the edge of the blade and insert the knife again. Cut through again, this time withdrawing the blade at the opposite angle, until you have fashioned the pocket. Stuff and roast.

STEW

Cut all the way through the neck and then cut the meat off into chunks approximately 2 inches square. Use in your favorite stew recipe.

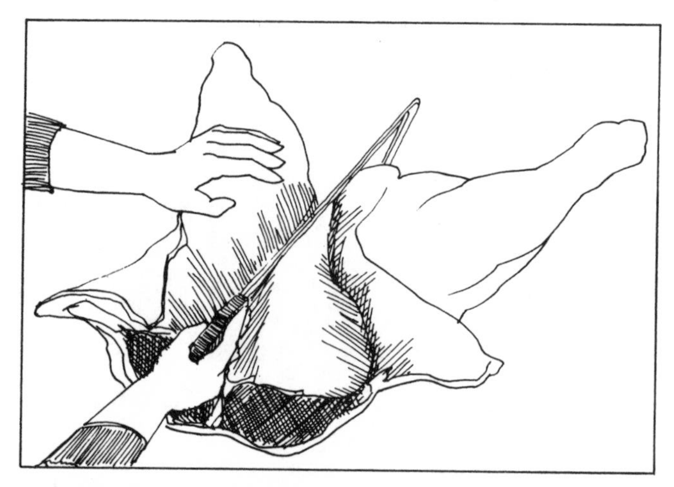

V-34. SPLITTING HINDSADDLE

THE HINDSADDLE

If you purchase an entire hindsaddle of veal, both hind legs will be attached; the piece consists of the entire rear half of the veal. You can cut this in two by sawing directly through the center of the backbone, horizontally, until you have split the hindsaddle in two (V–34), finishing off the split with the knife. (You may also purchase one hindquarter half—in which case, start with this.)

From the hindquarter you have the loin, the leg and rump and the shank. In our way of cutting this piece, the loin is separated from the rump and leg, and the sirloin portion is part of the rump and leg section.

LOIN

When working with the hindquarter, the first step is to separate the loin veal-chop section from the rest of the hindquarter. Place the hindquarter skin side up on your cutting area. Rub your thumb along from the chop end until you reach the top of the hipbone. When your thumb comes up against the hipbone, make an incision, first marking it with a knife (V–35). Then use the saw to cut straight down to cut off this entire section.

Now place the veal chop section (the loin section) skin side down. Open it up so that the edge of the skin will flap down. Using the knife, remove the kidneys from this section. They will be encased in a large clump of fat, which will sometimes make them invisible. The backbone of this section will be facing *away* from you (V–36).

To remove the flank, measure approximately 6 inches from the eye of the chop and make a small mark to guide you. Lay the piece on its back, holding the flap of skin firmly in one hand (V–37). Look at both sides of the chop ends. This next incision is a little involved when you first do it. Starting at the side away from you, at the bone end some 4 to 6 inches from the eye of the chop, make another incision, first with the saw, then continuing with the knife; the incision runs at a downward angle, so that it will end up at the eye of the chop at the other end of the piece. *Caution:* Do not cut through the chop at the other end. Line up both ends to avoid cutting through the eye.

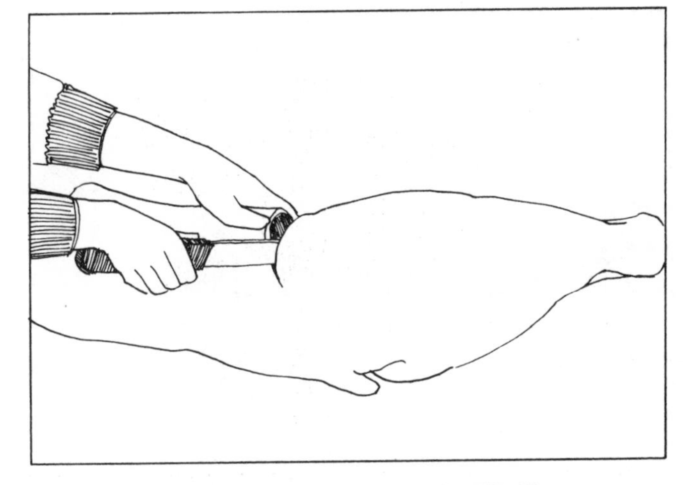

V-35. FINDING HIP; MARKING AND CUTTING THROUGH TO REMOVE LOIN

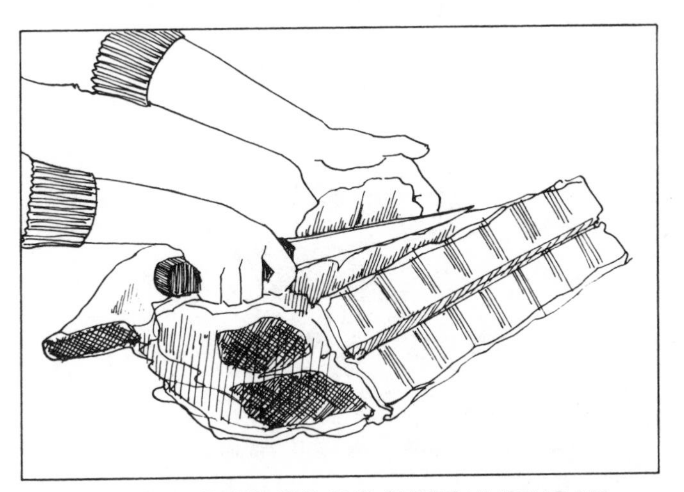

V-36. REMOVING KIDNEYS AND SURROUNDING FAT

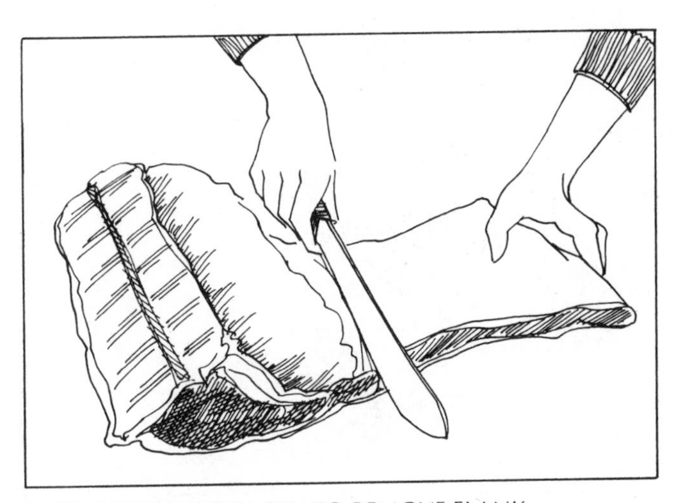

V-37. POSITION OF LOIN TO REMOVE FLANK

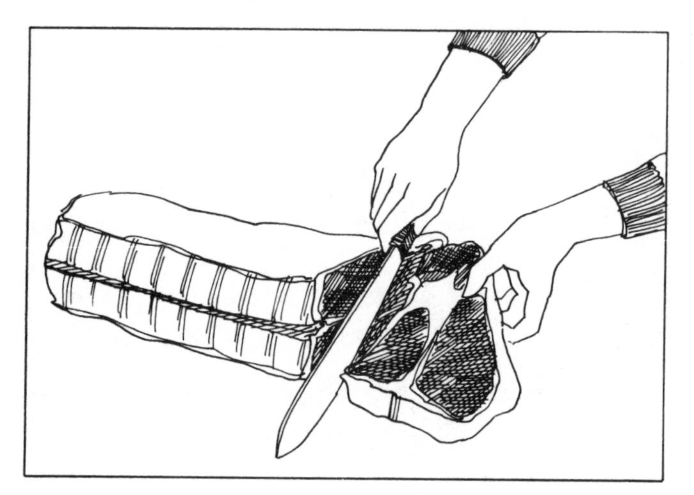

V–38. CUTTING LOIN CHOPS

The flat flank is now separated from the half-rack. Remove fat, bone, and membranes from flank, and set aside for use in stock or grinding. The remaining piece is a half-rack of loin chops. Scoop out the remaining large piece of fat on the underside of the half-rack.

LOIN VEAL CHOPS

Place the half-rack on the cutting area with the skin side up, underside down. Cut your chops to desired thickness, using a 10- or 12-inch knife, and cutting down into the meat as far as you can go. The meat lies in a half-circle over the T-section of the bone. When you reach this T-section, change to the saw and saw through the bone entirely. Go back to the knife to cleanly finish the last of the cutting (V–38). Trim off any excess fat from each chop (V–39).

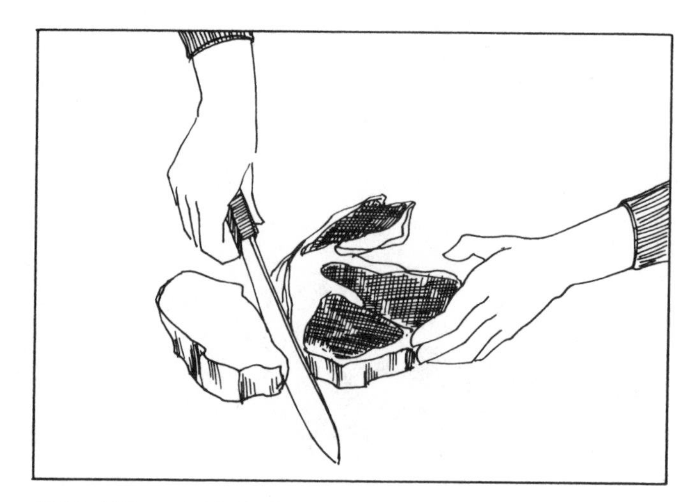

V–39. TRIMMING CHOPS

STUFFED DOUBLE VEAL CHOPS

Cut each chop, as described above, approximately 2 inches thick (V–40). To make a pocket in a double-thick chop, insert the point of a 6-inch blade in the outer edge of the chop (V–41). Push until you hit the T-bone in the chop. Work the blade point to the right and to the left to enlarge pocket, withdraw the knife, and reverse the blade edge. Reinsert the knife in same incision and repeat left-and-right motion to enlarge the pocket. Withdraw the blade, and the pocket is ready for stuffing.

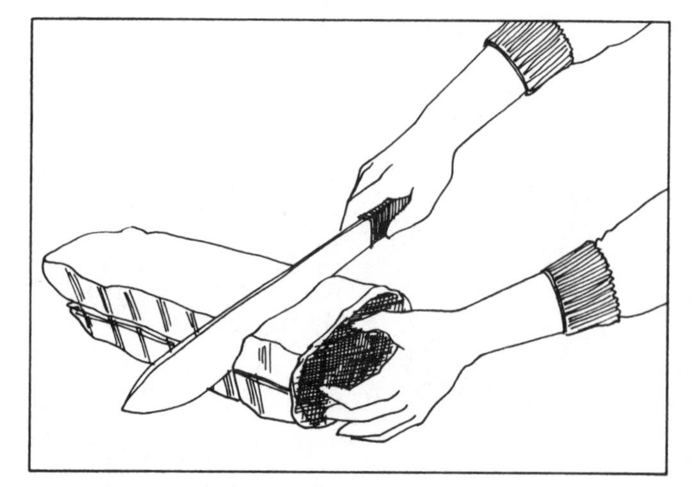

V–40. CUTTING DOUBLE CHOPS

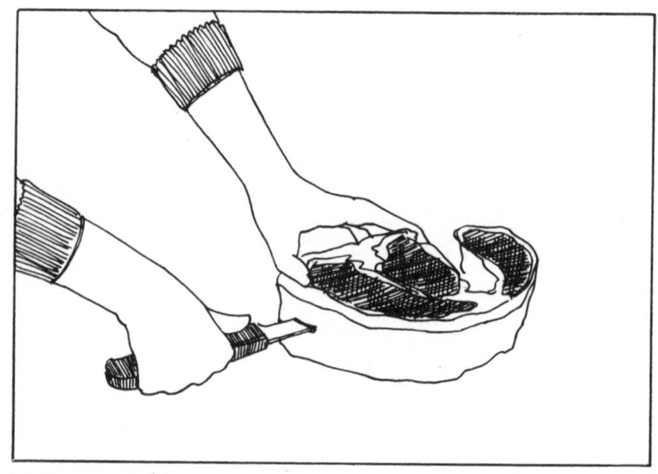

V–41. MAKING POCKET FOR STUFFING DOUBLE CHOP

SINGLE VEAL KIDNEY CHOPS

Cut off a 1-inch-thick piece of a veal kidney. Insert the kidney slice into a single veal chop by taking the tail of the chop and curling it around the kidney piece. Tie with string to hold in place.

BONELESS LOIN ROAST

We do not advocate the ordinary veal loin roast, which is often listed as a cut, because we feel it is not a good piece for roasting. The bones must be cracked so deeply that they are almost chops. This meat, boned, makes a much superior roast. Lay the half-rack piece skin side down with the backbone away from you. Remove the large piece of fat (V–42). Using the point of a 6-inch blade, cut down along the inner side of the backbone and around the T-bone. You will be cutting *around* the round fillet portion of the meat, thick at the front part and thinner at the other end. Loosen the meat up into the area where the T-bone ends (V–43).

Sometimes you will see 2 rib bones at one end of the loin. Slide the knife *under* the meat along the rib bones (V–44) and loosen the meat from the rib bones. Remove bones entirely. The backbone still remains. Turn the piece to stand on the backbone and pull the round fillet portion of the meat back with your hand to make the T-bone section clearly visible. Glide your blade down the inner (meat) side of the T-bone, staying close to the bone (V–45). At the bottom you will note a small enlargement of bone on

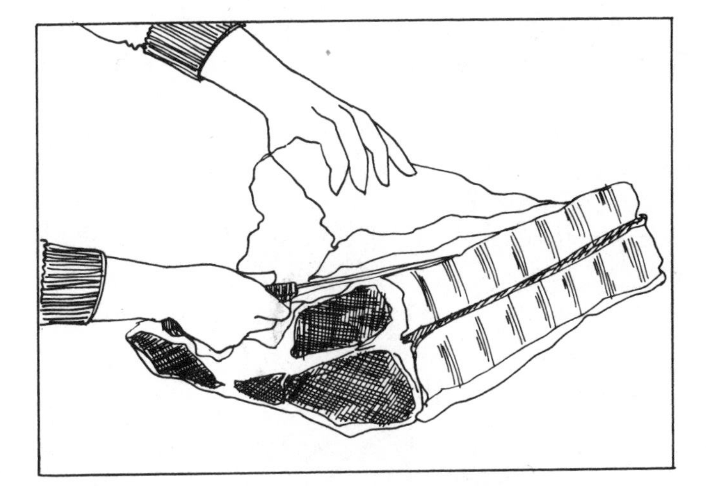

V–42. REMOVING FAT FROM INSIDE OF LOIN

V–43. LOOSENING MEAT IN T-BONE AREA

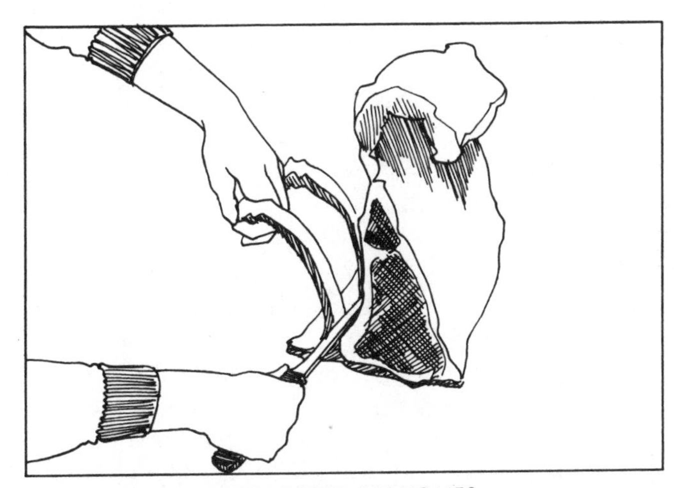

V–44. SLIDING KNIFE UNDER RIB BONES

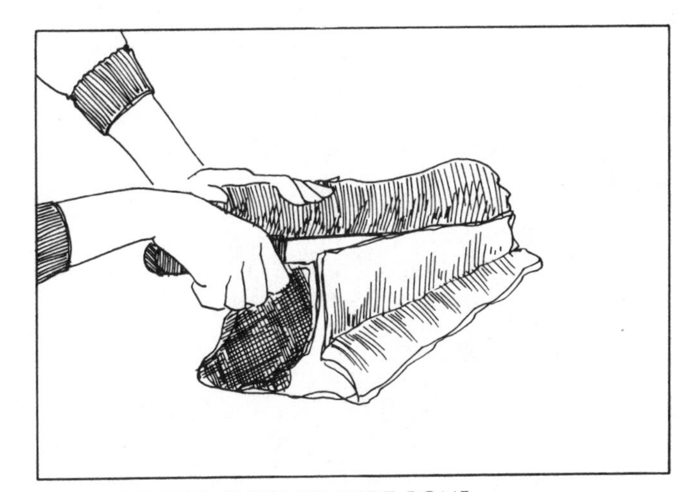

V–45. CUTTING CLOSE TO THE T-BONE

the inner side of the backbone. Trim around this with the knife until everything is cut away and the bone comes off cleanly.

You now have a boneless loin of veal. Trim away excess fat. The boneless loin can be stuffed and rolled, or the whole kidney inserted in place of stuffing. Lay the piece out flat, skin side down, take edge of flap and roll horizontally and tie (V–46). The bones may be used for stock or roasted with the veal.

BONELESS LOIN VEAL CHOPS

Tie the boneless loin as prepared above, then cut into chops, cutting in between and parallel to the strings.

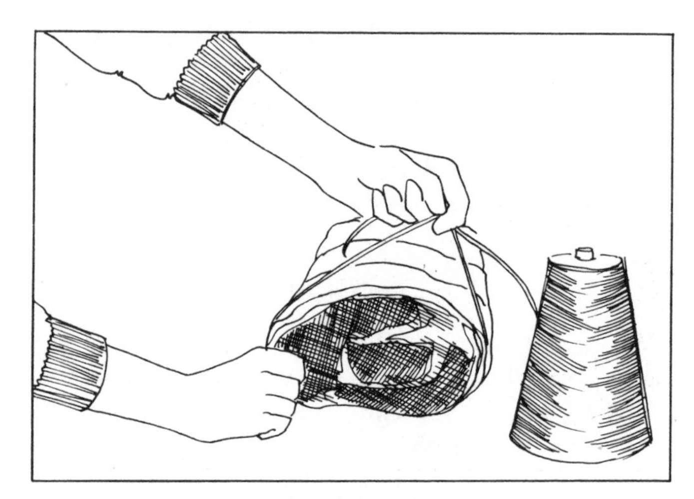

V–46. ROLLED BONELESS LOIN FOR TYING

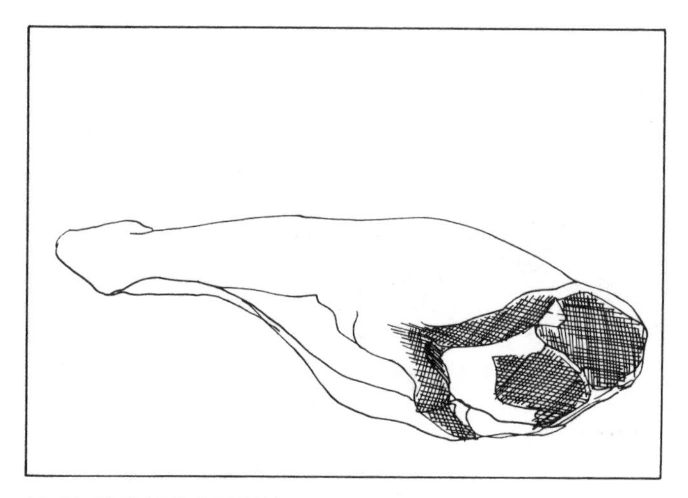

V–47. THE LEG OF VEAL

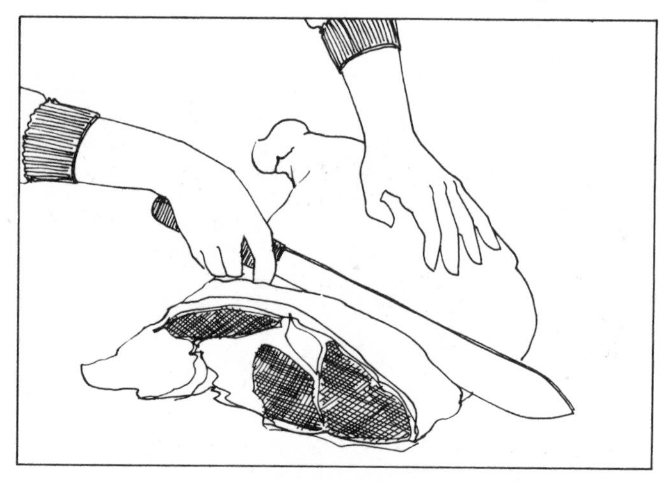

V–48. FIRST CUT: HIPBONE CHOP

LEG AND RUMP

To utilize this large section (V–47) in the best manner, it must be separated into smaller pieces.

PIN OR HIPBONE VEAL CHOPS

Measure some 2 inches from the hip end of the rump and, using a 12-inch knife, cut straight across and down to the bone (V–48). Use a saw to cut through the end of the hipbone. You will now have a large flat chop, some 2 inches wide. Trim all excess fat away. Stand on end to bake, lay on a bed of vegetables to braise, or broil.

In our cutting system, the sirloin portion of the hindquarter will be left on the rump and leg section.

Place remaining rump and leg section with *thin* skin side up. (The other side will have thick skin and fat.) Slide blade of a 6-inch knife under the skin and remove the thin outer skin and the outer membrane just under it (V–49). Do not try to remove the outer skin in one piece, but do it in sections, turning the piece as you work for convenience. When you have finished, the salmon-pink meat will be facing you.

Turn the piece over and remove skin, fat, and membrane from the other side (bottom). There will be more fat on this side. Keep the knife flat, cutting and gliding just under the skin and pulling the skin off by hand as you go. Use the tip of the blade to lift up the outer membrane (V–50). Then slide the blade flat beneath it to cut it away (V–51). Trim this side

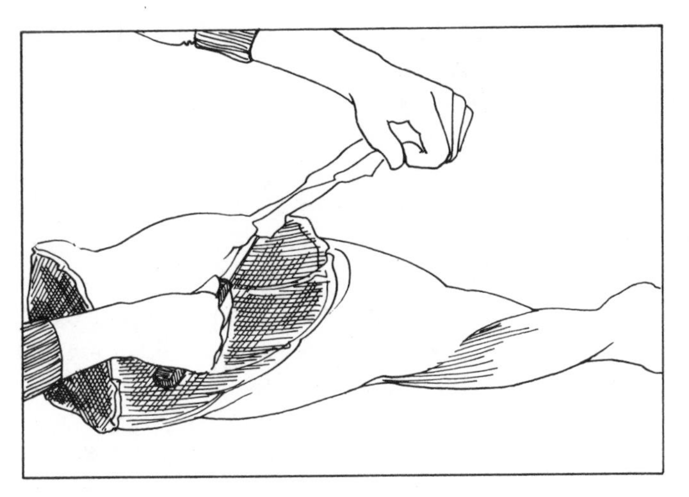

V–49. REMOVING THIN OUTER SKIN

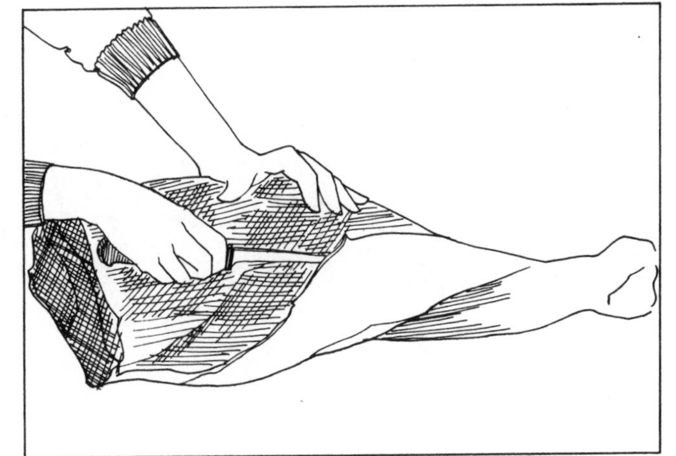

V–50. USING TIP OF BLADE TO LIFT SKIN

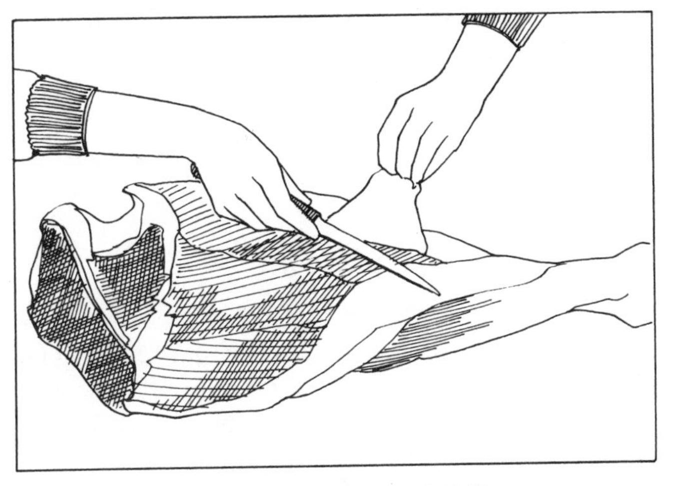

V–51. SLIDING BLADE UNDER MEMBRANE

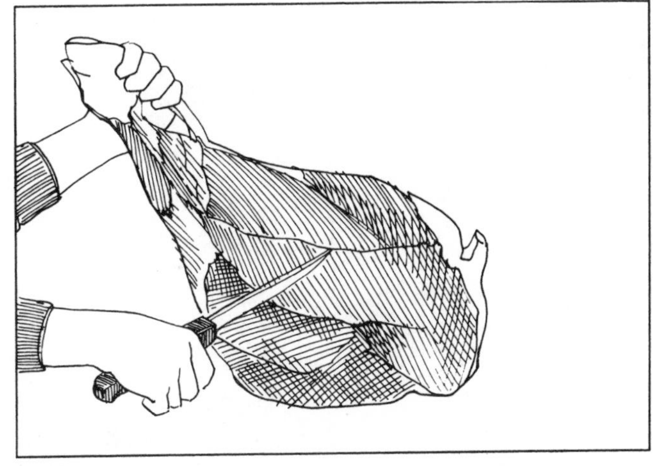

V–52. VARIOUS SECTIONS OF SALMON-PINK MEAT ARE NOW VISIBLE

entirely until only the salmon-pink meat is left (V–52). At the hip end there will be a large section of fat which should be removed to expose the slightly curved hipbone.

Go down to the leg (shank) and remove all the outer skin and fat. Take the tendon in hand and cut it off. One end will still be attached to the bottom of the hock. Completely remove the bottom of the hock and the tendon by cutting off this end of the leg with the saw just above the hock where the meat of the leg begins. Trim the thin outer fat from the leg bone.

The piece is now trimmed for further cutting. Turn it so that the underside is down again. There will be *seams* visible in the meat. Follow the seams in cutting. Starting with the largest seam at the top of the piece (V–53), use the front part of the knife to cut into these seams with small, short, probing cuts, following the seams as deep as they go.

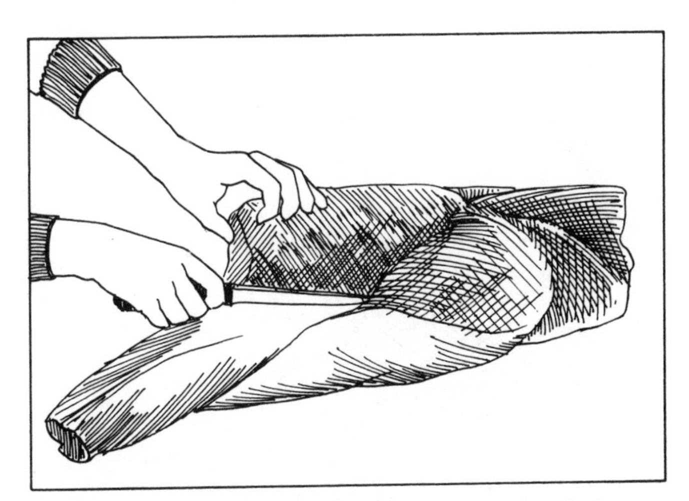

V–53. LOCATE LARGEST SEAM AT TOP AND REMOVE WHOLE SECTION (THIS IS THE "BOTTOM ROUND" SECTION)

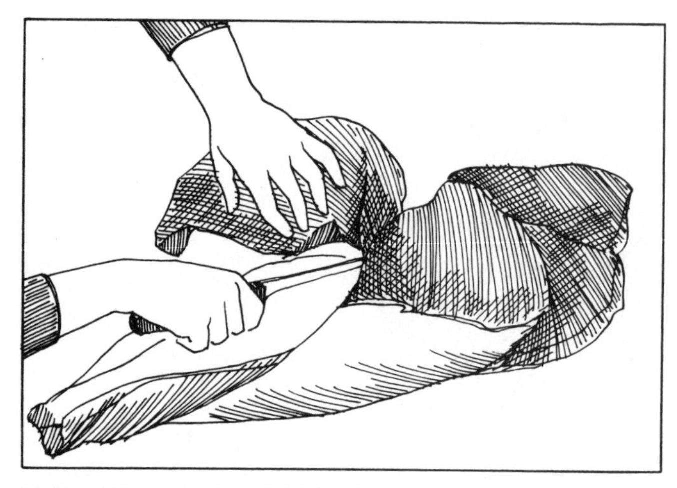

V-54. USING HAND, PULL BACK ON MEAT AS SEAM IS FOLLOWED AND CUT

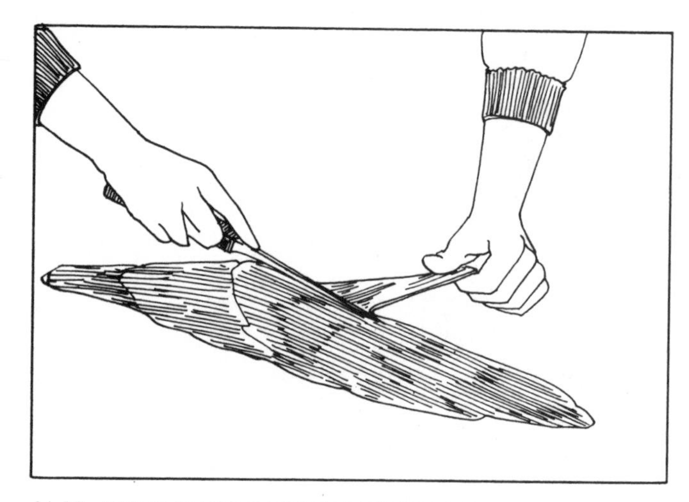

V-55. TRIMMING MEMBRANE FROM BOTTOM ROUND

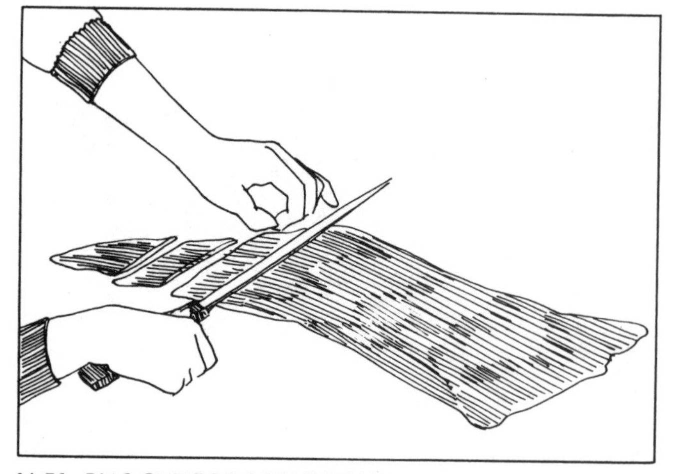

V-56. BIAS CUT FOR SCALOPPINE

As you cut, pull the meat back with the other hand. What you are doing is separating the sections of veal to give you smooth, seamless pieces when finished (V–54). As you make your small cuts and pull with the other hand, the seams will open up and the sections of meat will separate. As you do this, you'll come to the inner fat sections. Work around these with the tip of the knife and remove the fat. Continue to pull and cut, and the meat will separate on its own as you follow the seam lines. Soon you will be pulling this large piece of meat back in the direction in which you started. Continue with the small cuts until you have finished. The free section of meat is a bottom round (V–55).

Using a 5-inch paring knife, trim the piece you have removed, cleaning off all remaining outer membranes which may have come off with it. The bottom round piece is flat on one side. Place this flat side down. One end will be long and thin; the other end will be triangular in shape and the center a gentle mound. Note the way the grain runs in the meat.

SCALOPPINE

Using a very sharp 10- to 12-inch knife, turn the long, thin end of the meat toward you. Slice the first piece on the bias as you would London broil, against the grain. Continue to slice off thin pieces on the bias (V–56). These are the scallops for scaloppine. You can use the entire bottom round for scaloppine or cut off as much as you like and freeze the rest of the section *uncut.*

STUFFED ROAST BONELESS VEAL

Use the entire section of the bottom round. Simply cut a pocket in the whole section, stuff, and roast after tying.

EYE ROUND

The eye round is the next section to be freed. This is the large, cylindrical section that stands up next to where the bottom round had been (V–57). Use the tip of the blade and follow the natural line of the seam of the meat, cutting and pulling. As the eye round comes free you'll reach the other end of the hipbone. Cut around it, pulling the meat toward you, and continue cutting until you have cut away the en-

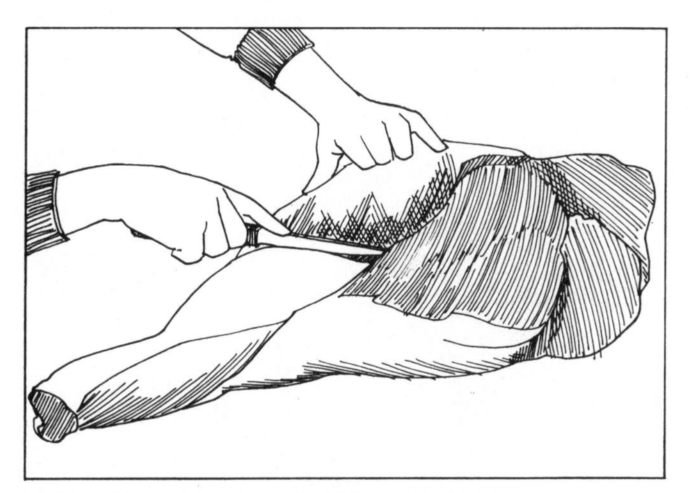

V–57. EYE ROUND CYLINDRICAL SECTION

V–58. "SEAMING OUT" THE EYE ROUND

tire eye round piece (V–58). Trim skin and outer fat from the eye round piece (V–59).

MEDAILLONS DE VEAU

Using the 6-inch blade, slice *straight down* for medaillons, making each slice ¼ to ½ inch thick (V–60).

ROAST EYE OF VEAL

Take the entire eye round section you have just removed and trimmed, season to taste, baste with a simple butter sauce, and roast in a 350° oven for approximately 45 minutes to an hour.

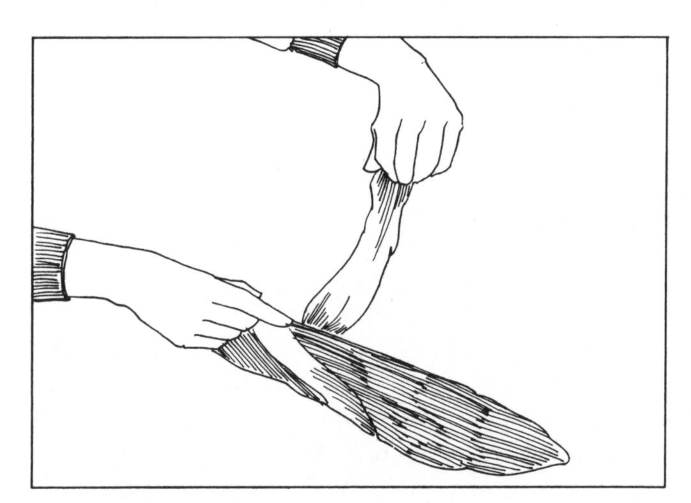

V–59. TRIMMING AND SKINNING

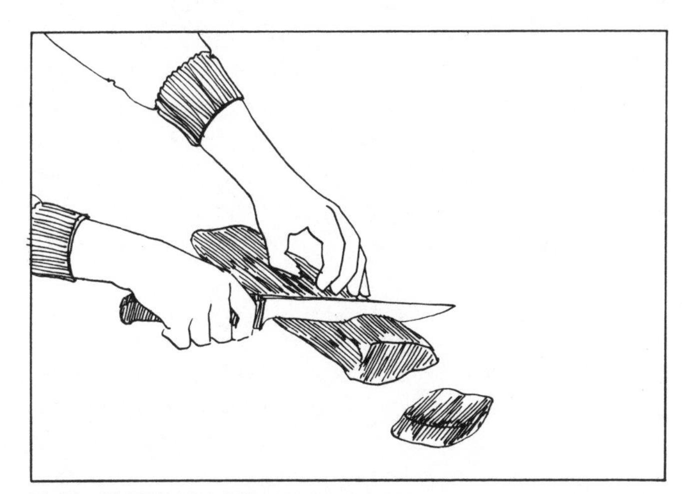

V–60. CUTTING *MEDAILLONS DE VEAU*

SHORT SIRLOIN, FLANK, AND TOP SIRLOIN

SHORT SIRLOIN

Look for the visible seam along the side of the piece of meat that remains. Cut and pull again in the same manner until you come to the hipbone. This piece, the short sirloin, lies *on top* of the hipbone. Cut it away and remove it entirely (V–61). You will have a flat, squarish piece. Trim all the membrane remaining on both sides.

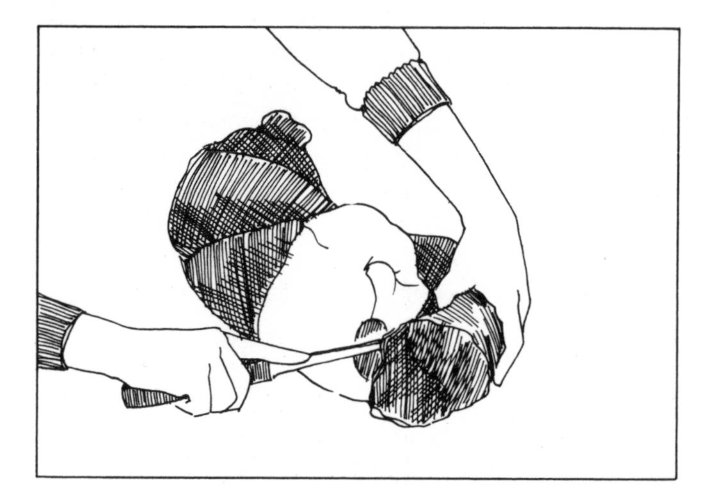

V–61. SHORT SIRLOIN ON TOP OF HIPBONE

VEAL KABOBS

Cut cubes from this short sirloin piece into 2-inch chunks.

VEAL FONDUE

Cut into ½-inch cubes.

SCALOPPINE

Scaloppine may also be cut from this section. Cut them as described earlier.

FLANK AND TOP SIRLOIN

Remove the small section of flank that lies on the side of the piece, once again following the natural seams in the meat (V–62). Set this aside.

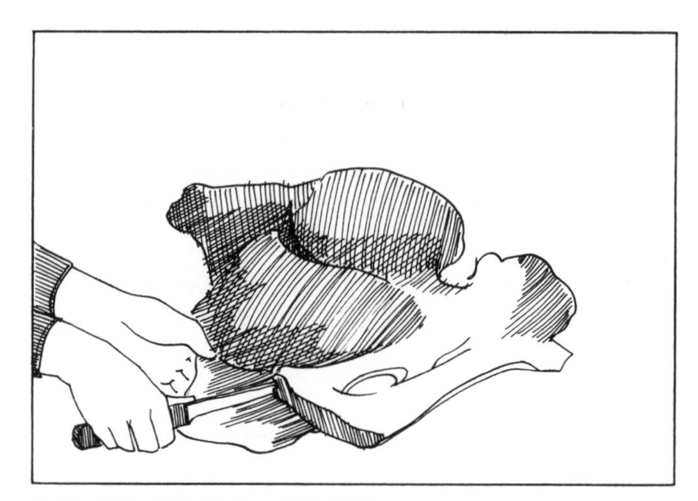

V–62. REMOVING FLANK PIECE

The next large section is the top sirloin, which is alongside the place where the flank was removed and lies on top of the piece of bone that extends directly from the leg. Use the 6-inch knife to free this piece. Locate the bone underneath with the point of the blade and use the point to cut first along the top of the bone and then around the bone. As you cut and pull around the bone, the leg bone will become exposed (V–63). Continue cutting and pulling around the bone, still following the seam, reaching the round end of the knucklebone. Continue to cut around this joint, cutting and pulling down into the seam until the entire piece comes free. This is the top sirloin.

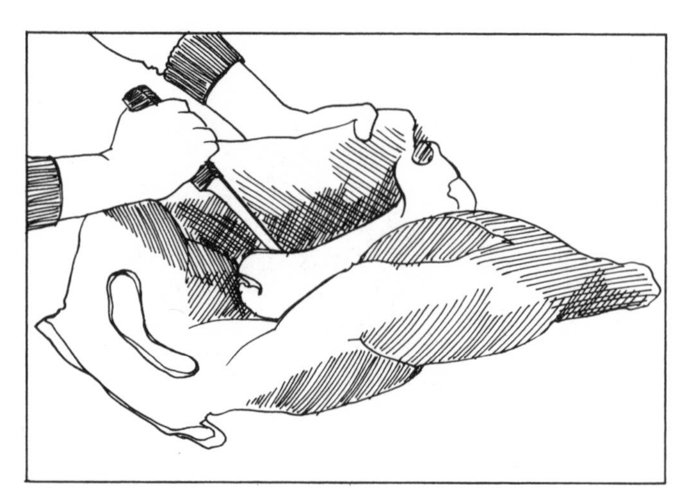

V–63. EXPOSING BONE

ROAST SIRLOIN OF VEAL

Roast this piece according to your recipe. Some people lard this section; we do not see the need to lard veal when it is carefully roasted because the desirable fat-free quality of veal is lost when it is injected with fat.

TOP ROUND

The top round of veal is next. Turn over the piece so that the entire hipbone section is on top facing you, along with the attached section of the backbone. Using your 6-inch knife, start at the backbone and cut around the hipbone to remove the bone from the meat (V–64). You will be cutting around the very shiny, smooth ball of the joint along with the connecting sinew and cartilege. The bone section should now wiggle freely. Continue to cut away until the entire hipbone with the backbone section comes free. The smooth, shiny round ball-and-socket joint will still be in your main piece of meat, however (V–65).

The top round is an easily movable, soft, large piece. Also note the thin top flap that lies over the top round section, and a seam separating one edge of this piece. Cut and pull along this seam until you have pulled it back and off altogether. Set this aside with your flank piece.

The top round piece is not yet freed. To do this, find the seam that lies against the leg (shank). Cut and pull until the piece comes off (V–66); cut around the end of the leg bone and free the meat entirely.

Trim the outer membranes and fat from both sides of the top round piece (V–67).

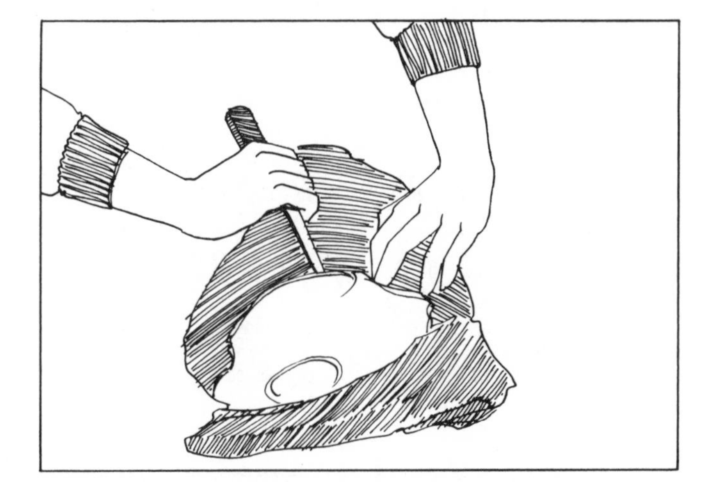

V–64. CUTTING AROUND HIPBONE TO REMOVE

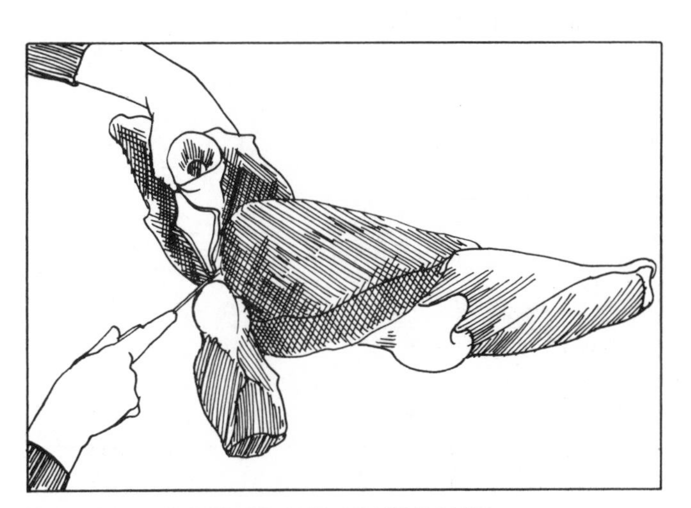

V–65. TAKING OFF HIP AND PELVIS BONES

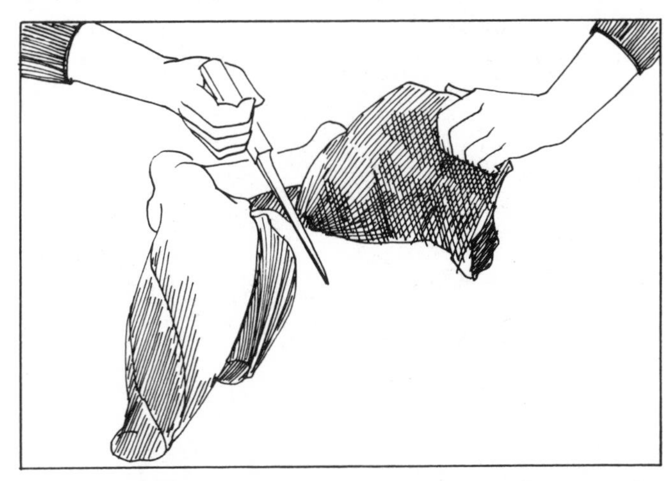

V–66. FREEING TOP ROUND BY CUTTING AROUND LEG BONE

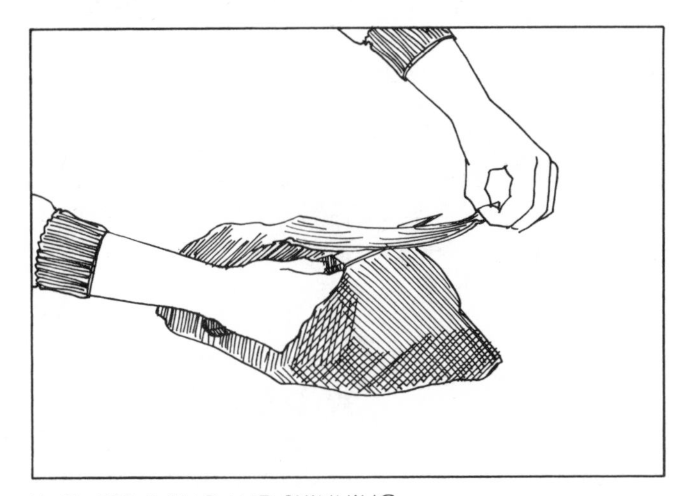

V–67. TRIMMING AND SKINNING

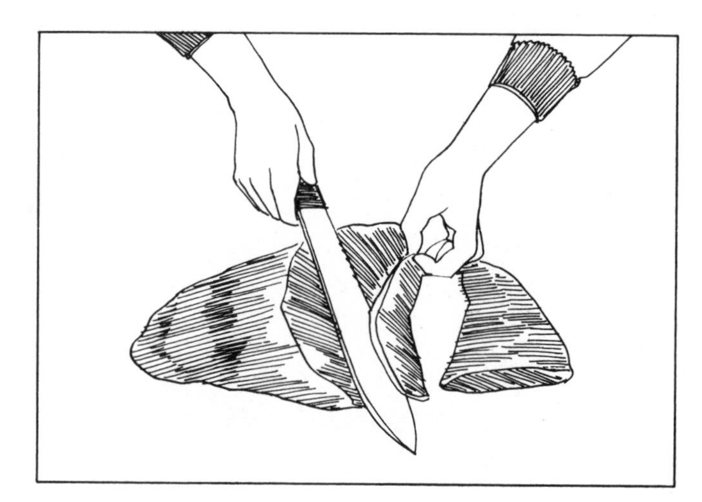

V-68. CUTTING CUTLETS FROM TOP ROUND

VEAL STEAK

Place the top round piece flat side down. The piece will have a natural bias shape to it on the short side at the top. Following this natural line, cut down into this, on the bias (V–68), to make approximately 1-inch-thick veal steaks. You should get about 5 veal steaks out of a top round of veal.

VEAL CUTLETS

Cut the same way as for veal steaks, following the same natural bias line, but cut them only ½ inch thick.

SCALOPPINE

Scaloppine may also be taken from the top round. Cut on bias, not more than ⅛ inch thick. Flatten and prepare as desired.

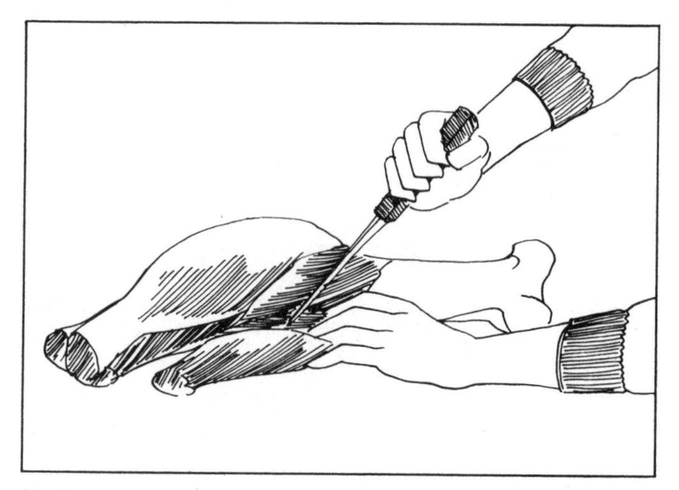

V-69. CUTTING MUSCLE OVER SHANK

SHANK

The shank is left. You will see a seam alongside it. Along this seam, cut this oblong piece of meat halfway through (V–69). Then turn the shank bone and place the ball-and-joint socket against your stomach for support. Finish removing the rest of the meat around the shank (V–70). Set aside with the flank and other pieces you've saved.

Move the hock joint back and forth and make an incision encircling the leg above the hock joint (below the meaty part of the leg). Now stand your shank on the end of the bone and curve it back at a right angle. Plunge the point of a paring knife straight into the joint (V–71) and work around the joint until you have removed the entire shank (V–72). Set the joint aside. The shank is ready for Osso Buco.

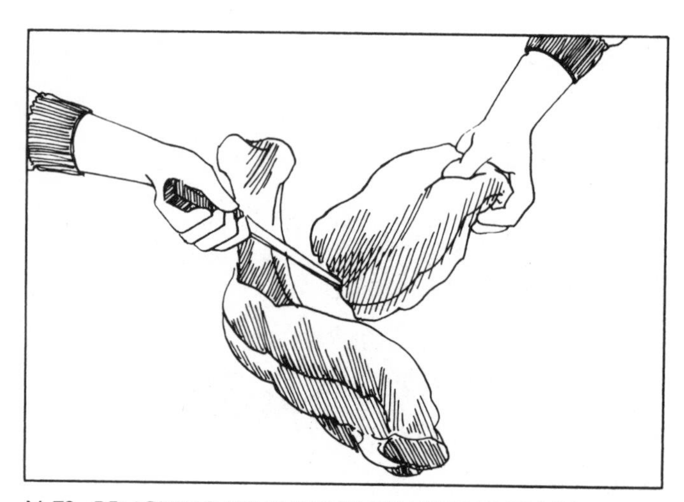

V-70. REMOVING REMAINDER OF MEAT AROUND SHANK

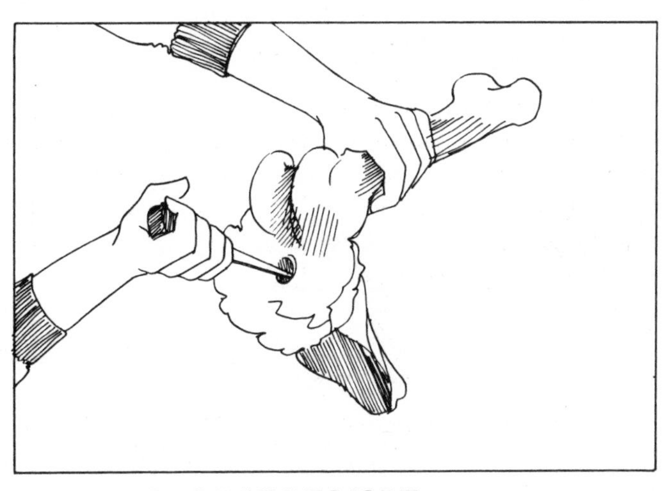

V-71. PLUNGING KNIFE INTO JOINT

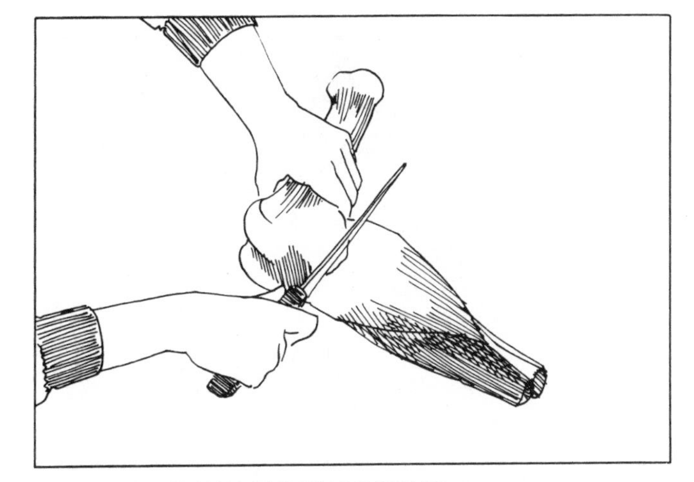

V-72. TAKING SHANK OFF LEG BONE

OSSO BUCO

This hind-shank section is the one we most recommend for good Osso Buco. Trim fat, muscle and membranes from the shank, leaving the meaty sections on. Using the saw, cut the shank through exactly in the center to make 2 pieces. In a commercial market the shank is often cut into 3 or 4 pieces; this should be done with a motorized band saw. If you have such a saw available, you can cut your shank in 4 pieces, but with the regular handsaw we suggest making only the single center cut. You can also use the entire leg and cut meat from it after it is cooked (V–73).

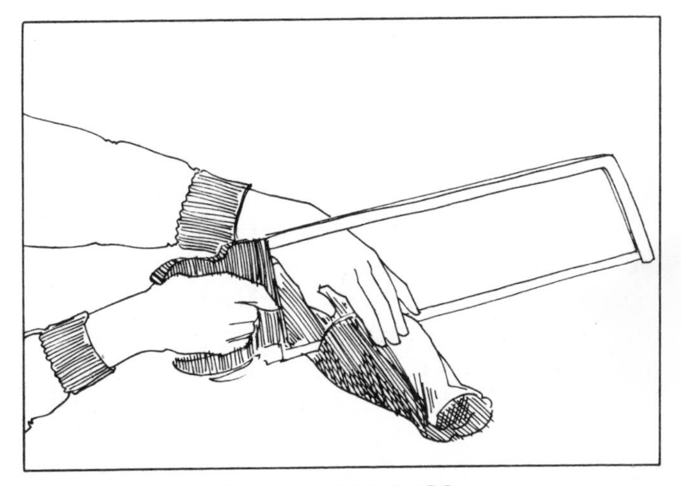

V-73. CUTTING PIECE FOR OSSO BUCO

SET-ASIDE PIECES

The pieces of meat you have set aside while cutting up the various veal sections (V–74) may all be used for any slow-cooking dishes, such as stews. They should be trimmed of excess fat and membranes before using. For Veal Stroganoff cut small cubes of meat. For Veal Stew Elegante with a delicate flavor, cut the meat into somewhat larger cubes. For vealburgers, grind the pieces and use for a different, subtle taste in any ground-meat dish.

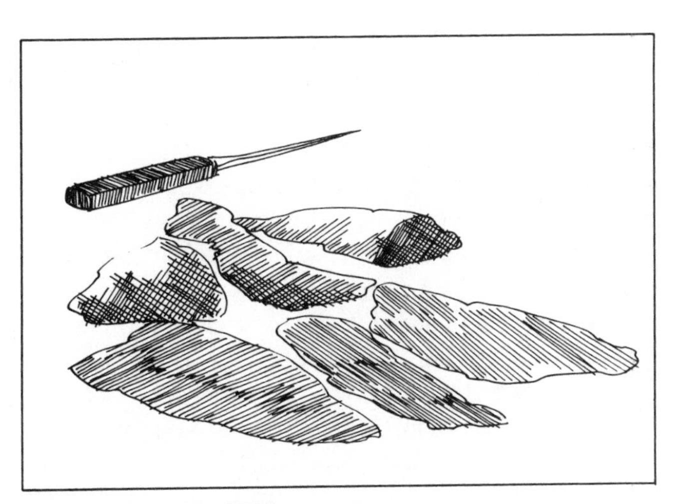

V-74. PIECES SET ASIDE

8

Beef

Americans consume almost as much beef and veal as they do all other kinds of meat together. The science of producing better, tastier beef with greater yield per carcass is of unending interest to American cattle producers. Numerous varieties of beef cattle reach the market. Ask which breeds your supplier is handling and, in time, you will develop your own preferences. The Black Angus (Aberdeen Angus) and the Hereford are probably the major types on the market. However, breeders have crossed beef cattle with Brahman bulls and Texas Longhorns to develop cattle more resistant to disease, better able to stand climatic changes, and capable of producing more meat. The Charolais, imported from France, is a breed gaining popularity as a beef producer and for breeding with American cattle. Some cattle breeders have produced herds called Brafords, cattle crossbred from Hereford and Brahman. The Brangus, which has been very successful, is another crossbred type out of the Brahman and Black Angus breeds.

When buying bulk meat, the most important factor to ask about is the diet the beef has been fed. We feel grain-fed Midwestern and Western beef still yields the most tasty meat. In choosing bulk purchases at the wholesale market, remember that prime beef is from steers that are from 1 to 2 years of age. Though practiced buyers can learn from the outer conformation of the animal, this requires many years of experience, and chances are your steers will already be halved or quartered. Remember, a thousand-pound steer dresses down to about 475 to 500 pounds of retail meat.

But there are ways to help yourself choose good meat. First, of course, there are the grading and yield marks discussed in Chapter 1, Buyer Tips. If you are purchasing a section of beef, the outside fat should be milky-white and fresh-looking to the eye. Avoid meat with yellow or deep-yellow outside fat. Even on an entire forequarter, you will be able to see the eye of the meat itself at the rib end. The meat should be amply grained, with a pink-red, alive tonality. Avoid meat that is dull or deep red and lusterless in tone. This indicates a lack of proper diet to produce the kind of beef you want. Inside the bones, the color should be a good, strong red, and between the bones, the meat should look bright and healthy. A pale color and the absence of a strong red in the bones indicate lack of quality and youth.

But here, as with retail buying, you will do best with a supplier whom you can trust and whose meat comes from good sources. Remember, even dressed-down beef is large and weighty. Make sure your facilities for handling and transporting match your ambitions. From the standpoint of pure economy, the best purchase would be an entire forequarter as a start.

Once again, we point out that you can purchase, at your bulk supplier's any of the major sections detailed here. You do not need to purchase an entire forequarter. For example, you can purchase separately the rib section or the plate section, the brisket or square-cut chuck. Follow our instructions for handling these sections specifically. But for those who have the facilities to avail themselves of the economy of the larger purchase, we begin with instructions on preparing the entire forequarter.

FOREQUARTER

The first step is to saw the forequarter into 2 manageable pieces. With the large 12-inch knife, make a cut right across the very center of the forequarter, from side to side. Then place the saw in the incision made by the knife and, holding onto one half of the piece with your hand for support, saw right down, with short motions, until you have sawed the piece in half (B–1).

One of these large halves is the rib and the plate; and the other consists of the cross-cut chuck with the brisket and shank (shoulder). Set the cross-cut chuck with the brisket and shank aside. We begin with the rib plate piece. This piece must be cut in two to separate the rib section from the plate.

Measure 12 inches from the bottom to the very center of the piece. Using a large knife, make an incision directly across the center from side to side (B–2). Position the saw in this incision and saw through the entire piece. The rib and plate (short plate) are now separated.

THE RIB

From this section, you get:

Rib of Beef
Standing Rib Roast (also called Beef Rib Roast)
Short Cut Rib Roast
Rib Steaks
Pepper Steaks
Boneless Ribs of Beef
Boneless Rib Steaks
Pot Roast
Spareribs (Beef)
Short Barbecue Ribs (also called Riblets; Finger Ribs)
Soup Meat

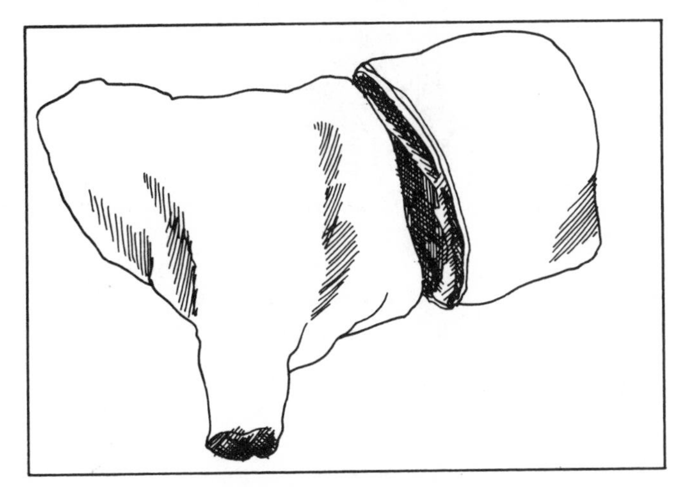

B–1. FOREQUARTER CUT INTO 2 MAIN PIECES

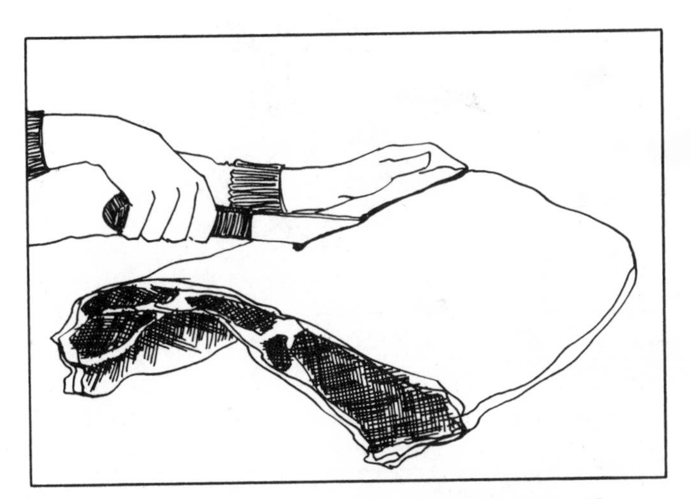

B–2. CUTTING DIRECTLY ACROSS CENTER RIB-PLATE SECTION

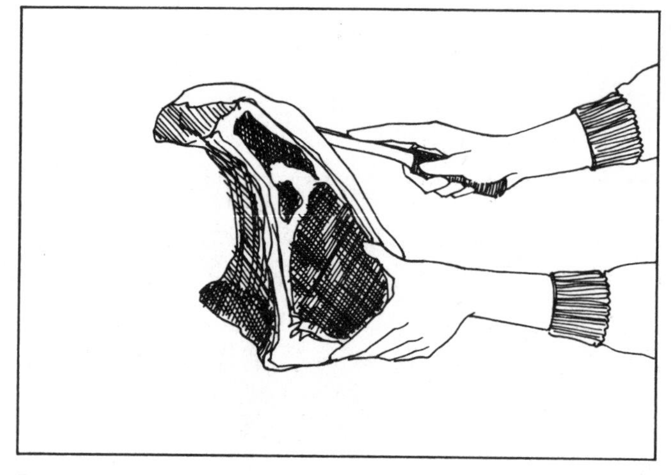

B-3. FOLLOWING LINE OF 3RD RIB

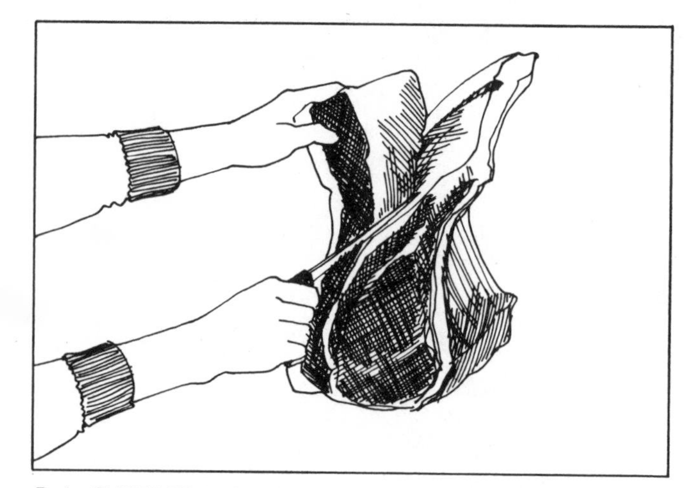

B-4. CUTTING ALONG NATURAL LINE OF FLAP TO REMOVE TOP OF RIB

Stand the rib piece on the thick chine bone. The ribs will be up into the air, curving away from you with the outer fat facing you. Look at the thinner end of the piece and find the single eye side of the meat. Count off 3 ribs from the eye of the meat at this end. Using your 6-inch knife, cut into the meat approximately ½ inch (B–3), following the line of the 3rd rib. Put your fingers in the cut and pull back the top layer of fat and the meat under it. With a knife again, cut farther in as you pull back. You will see a second layer of fat under this top fat-meat flap, which you must separate from the top layer as you pull.

Following the natural line of the flap (B–4), cut all the way to the bottom—you have now removed the top of the rib (B–5). Lay this piece flat and you will see a division where two pieces of fat come together with a line of meat between them (B–6). Cut down vertically to make two pieces out of this large flat section. Trim excess outer fat from these pieces. *Note:* Some people prefer to leave a little fat on them. You now have 2 pieces of meat that can be used for pot roast or soup meat, or cut thin for pepper steak.

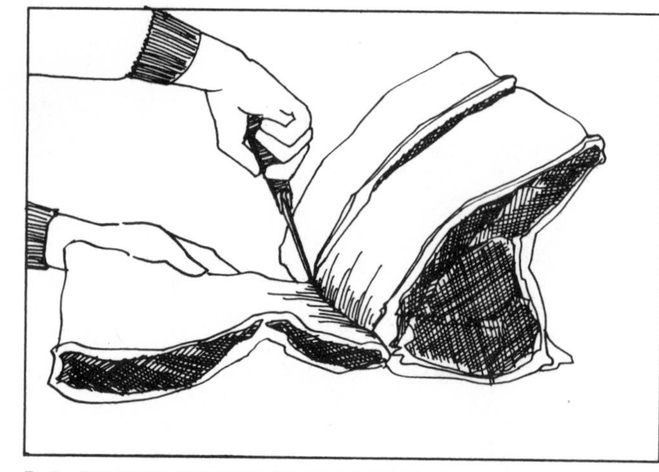

B-5. FINISHING SEPARATION OF THE 2 PIECES BY CUTTING ALONG BOTTOM

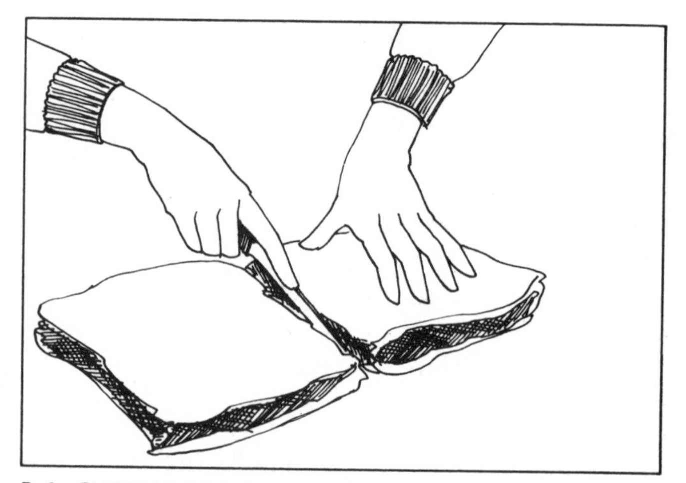

B-6. CUTTING TOP-OF-RIB PIECE

RIB OF BEEF

In cutting off the meat above, no bones have been removed. The piece remaining is a 7-bone rib of beef. One part has thicker outer fat because of the top of rib you have just cut away.

Turn the piece on its back, the outside fat down. Place the saw against the bottom of the ribs next to the chine bone and saw the rib bones at an angle (B–7) until you feel the saw has cut through them.

Change to the 8-inch knife, and, cutting into the incision you made with the saw, maneuver the knife close against the chine bone and around the meat, making small cuts. As convenient, turn the entire piece on the table as you cut, always keeping close to the chine bone. Cut around the entire chine bone, down its full length, and you will see the flat part of the backbone and the spinal cord. The spinal cord is a deep-yellow line. Follow this line with the knife, severing the cord. The chine bone and backbone will come free (B–8). The chine bone can be used as a roasting rack.

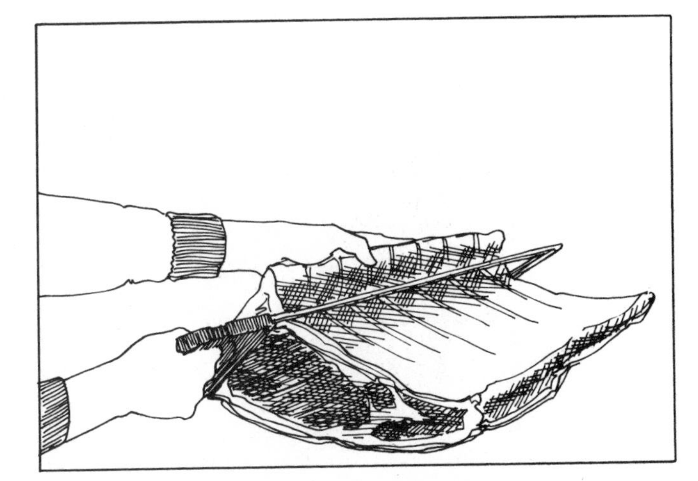
B–7. ANGLE FOR SAWING CHINE BONE

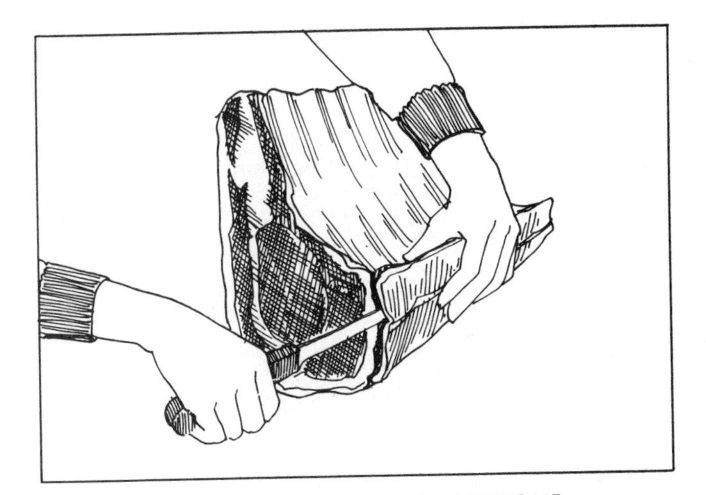
B–8. REMOVING CHINE BONE AND BACKBONE

STANDING RIB ROAST

You have just fashioned a 7-bone standing rib roast. This is a very large roast. For a smaller, average-size roast, measure off 4 ribs and cut down in between the 4th and 5th rib bones using your knife. *Note:* If the rib bones are too long for your oven or pan, cut the lower end of the rib bones off with the saw, all at once, sawing straight across. Set this oblong set of the cut rib ends aside for later use.

SHORT CUT RIB ROAST

This piece is exactly what the name says it is. Measure 1 inch from the eye of the meat along the rib bones and, with your saw, cut across all the ribs at once. When the ribs have been cut through, finish with the 12-inch knife. Trim off all the outer layer of fat. There will be a thin underlayer remaining. You'll also see remains of the deep-yellow spinal cord. Take this off entirely and you have a short cut rib roast.

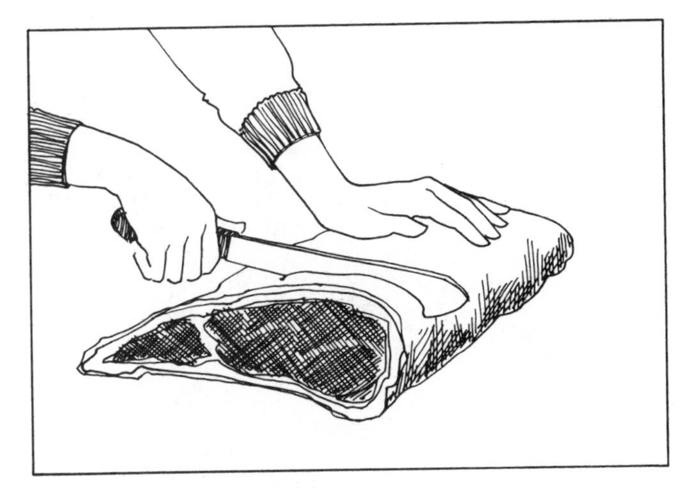
B–9. CUTTING RIB STEAKS

RIB STEAKS

You can make rib steaks from this shortened section by cutting between the ribs (B–9). For 1 steak, cut in front of the first rib bone. For 2 steaks, the cut

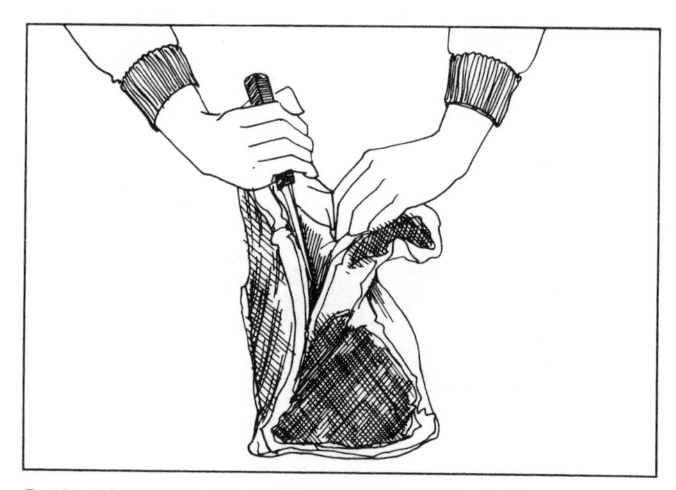
B-10. POSITION OF KNIFE TO BEGIN BONING

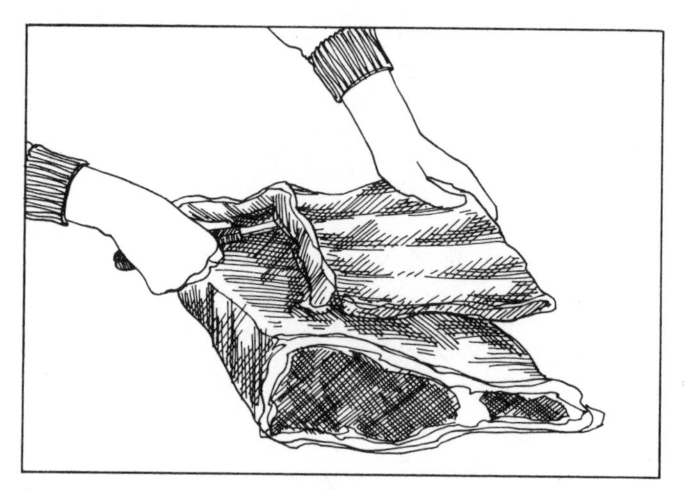
B-11. CUTTING ALONG NUBS OF BONE

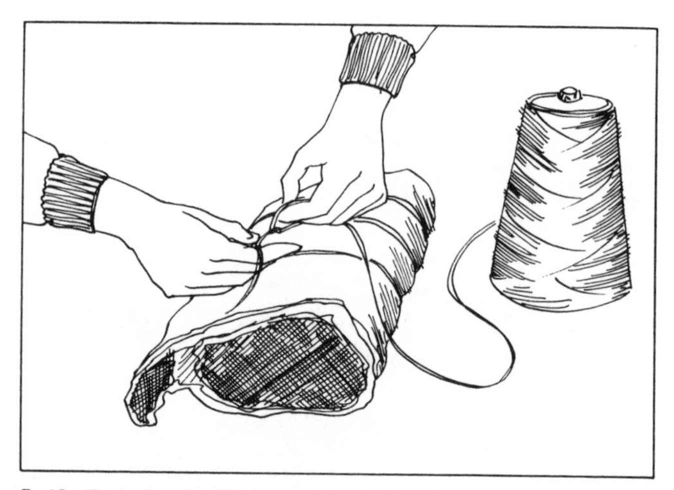
B-12. TYING BONED PIECE FOR ROASTING

is made against the opposite side of the bone. For 3 steaks, the 3rd cut is made right before the 2nd rib; 4 steaks, the next cut, is on the opposite side of the 2nd rib bone, etc.

BONELESS RIBS OF BEEF (BONELESS RIB EYE)

Place the rib roast on its thick end, with the cut bones facing up. To bone, hold the knife blade toward you and place the point of the knife into the top of the meat as close against the rib bones as you can (B–10). Start cutting down, working the blade close beside the bones. Keep cutting down alongside the bone until you come to little nubs of bone on the underside. Follow the contour of these bone nubs until all the rib bones are freed (B–11). Set aside. (These are regular spareribs.) You now have boneless ribs of beef.

To tie the boned piece, start at either end by making ties 2 inches apart. Then tie in between to make final ties 1 inch apart (B–12).

BONELESS RIB STEAKS

Slice the boned ribs to desired thickness. We suggest slicing after tying so that the boneless steaks keep their shape better. *Note:* If not from aged, quality meat, rib steaks can be very chewy. In that case, we suggest using only the first cut for rib steaks and the remainder for a boneless roast.

SPARERIBS

These are the ribs you removed in boning. Trim the excess fat off both sides, top and bottom, and they are ready for use.

SHORT BARBECUE RIBS

When cutting the standing rib roast, you may have sawed the long rib bones off to fit the roast into your pan or oven. The piece of short ribs left over is short barbecue spareribs, or bite-sized riblets. Trim the excess fat from both sides with the 6-inch knife, and they are ready to use for small spareribs, hors d'oeuvres, etc.

PLATE

This is the other half of the rib-plate section. From the plate, cut:

Skirt Steaks
Skirt Steak Rolls
Long Bone Flanken
Plate Flanken
Boneless Plate (also called Rolled Plate; Yankee Pot Roast)

Put the entire plate section fat side down, bone (bottom) side up. Begin ripping off the outer top skin, which is a fibrous layer (B–13). You will then see a flap of meat running diagonally across the entire plate section. Insert the knife at the top end of this flap facing you and work it into the seam of the flap.

Use the tip of the knife as you work into the flap, holding the top with your other hand (B–14). Pull the flap back by hand as you cut into the seam to remove the entire piece.

When this is removed it will be a long, thin strip of meat, the skirt steak (B–15), with some fat on it. Remove this fat (B–16).

SKIRT STEAKS

After removing the fat, turn the piece over. You will see a thin skin on the underside. You can usually pull this off with your hands. Use the knife to cut away any heavier fat that remains (B–16). When trimmed, cut the piece into thirds for skirt steaks.

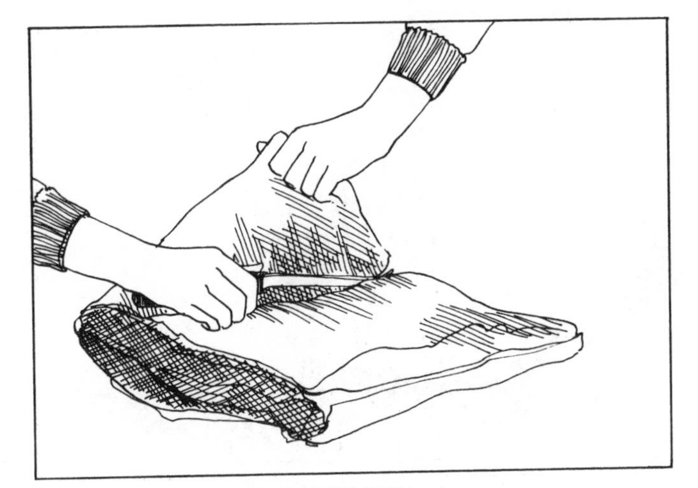

B-13. TRIMMING OFF OUTER SKIN

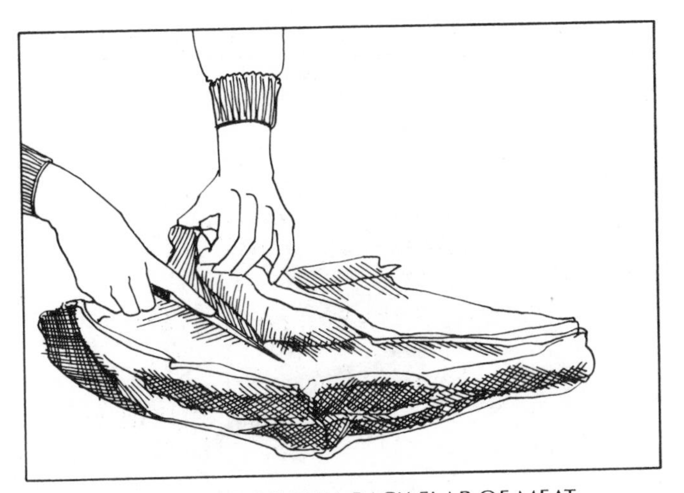

B-14. USING HAND TO PULL BACK FLAP OF MEAT

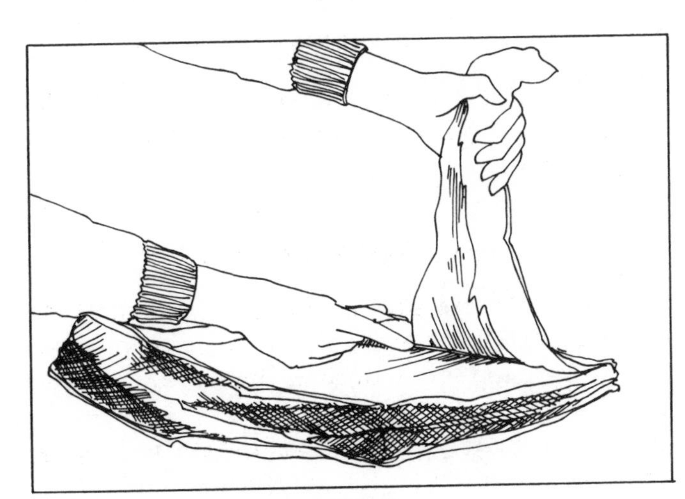

B-15. CUTTING OFF SKIRT STEAK

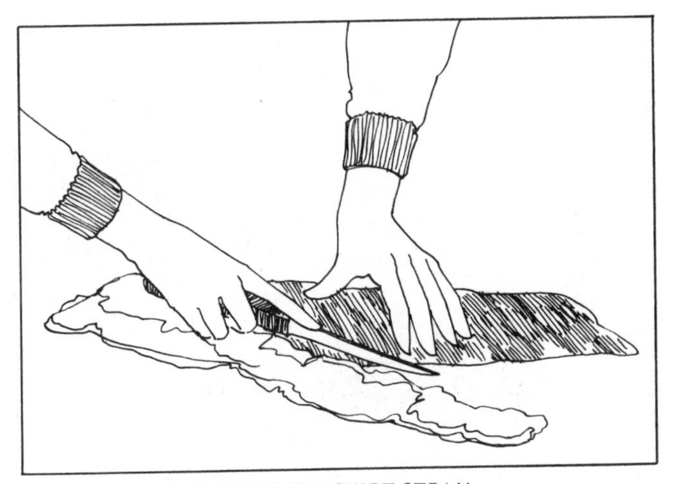

B-16. TRIMMING FAT FROM SKIRT STEAK

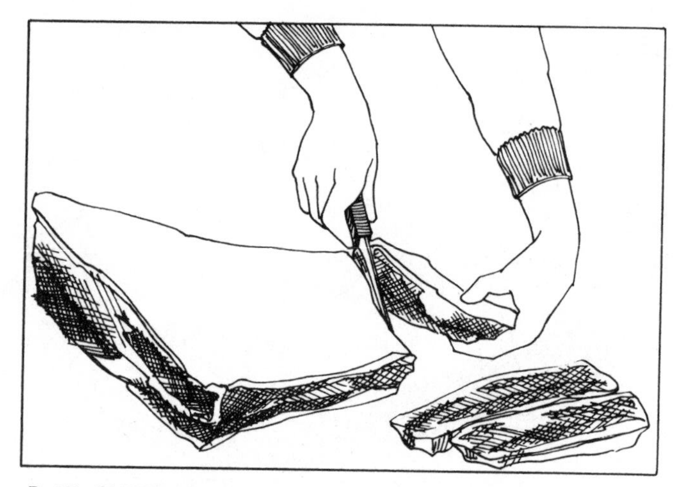

B-17. CUTTING LONG-BONE FLANKEN

SKIRT STEAK ROLLS

Cut the long, thin piece of skirt lengthwise into narrow strips, trimming off any ragged edges. Roll the narrow strips into small pinwheels of meat and fasten with a toothpick or skewer for boneless skirt rolls.

Now remove the upper skirt steak from the plate. You'll see this long strip lying at the top edge of the plate, looking very much like the lower skirt steak piece you just cut away. Holding the knife flat, cut under this top long piece, using your other hand to pull it back as you cut, until you have entirely removed it. Trim off the fat. *Note:* This piece is not to be used for skirt steak, but only for ground meat.

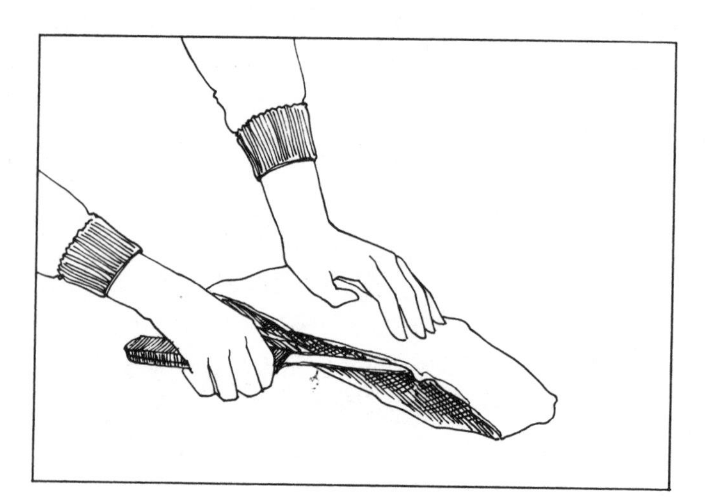

B-18. TRIMMING OUT FAT

LONG BONE FLANKEN

The main piece of plate is left. Find the 3 long horizontal bones (plate bones) from the shorter end of the piece. With a 12-inch knife, cut in between the bones (B–17) until you have 3 pieces each about 2½ inches wide. Use the cleaver here if necessary.

Trim the outer fat from each piece on both sides (B–18 and B–19). You will also see fat running through the center of each piece. Take out this fat by going in at one end of each piece with the knife and reaming out the fat with short cuts. When you have fully trimmed all fat and membranes from each piece, take the saw and cut down in the center of each piece into the bone. Saw until you feel the bone crack (B–20), approximately ¾ of the way through, and then remove saw. These are the long bone flanken.

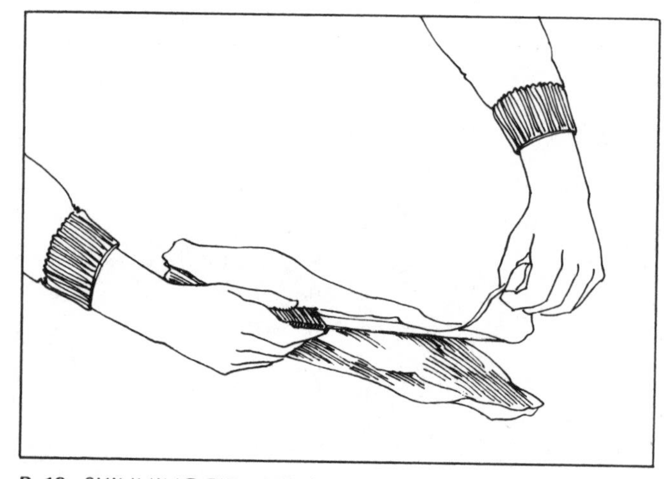

B-19. SKINNING FAT AND SINEW

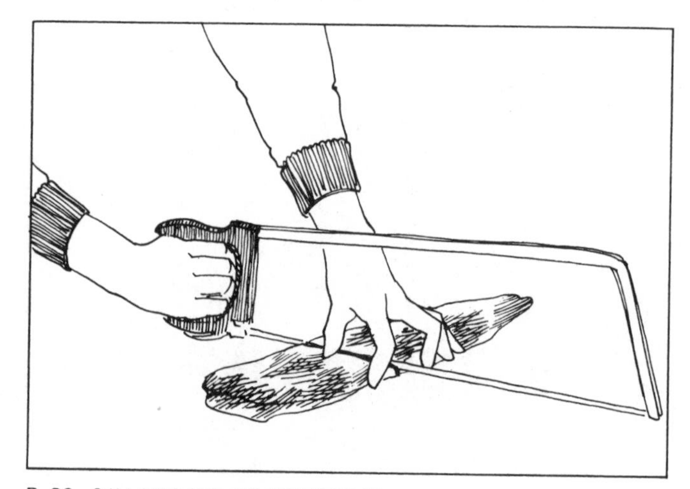

B-20. SAWING TO CRACK BONE

PLATE FLANKEN

Approximately half of the plate remains. Using the saw, cut the plate approximately in half, perpendicular to the bone (B–21). Finish the cut with the 12-inch knife. The square piece is the plate flanken.

Stand it on end and trim off the outer fat in long swoops, using the 12-inch knife (B–22). Lay the piece flat again, fat or outer side down, and saw into thirds through the bone (B–23). But saw only through the *bone,* not all the way through, and finish cutting through with your knife. Remove all the remaining outer fat down to the meat.

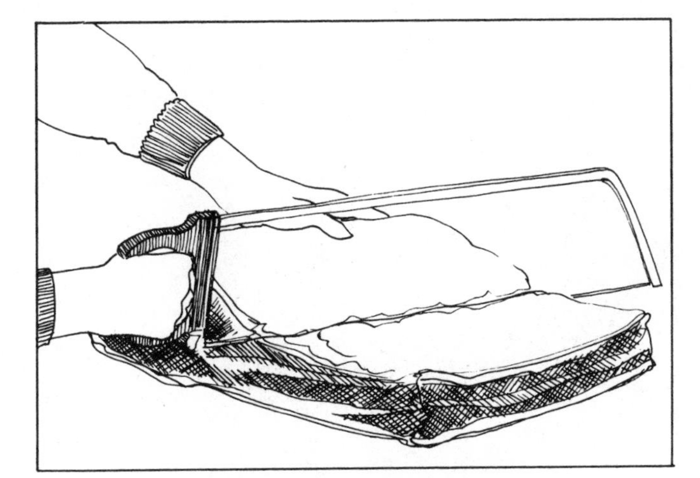

B-21. CUTTING PLATE IN HALF FOR FLANKEN

BONELESS PLATE (YANKEE POT ROAST)

Take the remaining piece of plate and stand it on the bone edge. There will be a flat piece of bone at one end. Place the knife against the edge of the bone and start working with the knife flat against the bone, pulling the flap of meat back with your other hand (B–24).

When you are halfway through, stand the entire piece on end and, holding it firmly with one hand and pressing it against your stomach, cut down with the knife around the bone, again pulling the meat toward you with your other hand. When you've removed bone, set aside.

Lay the other piece flat and cut it into a square shape. Then remove the outer fat down to the meat. Turn the piece over and do the same with the fat on the other side. Use both the 12-inch and 6-inch knives as convenient, working around the meat to

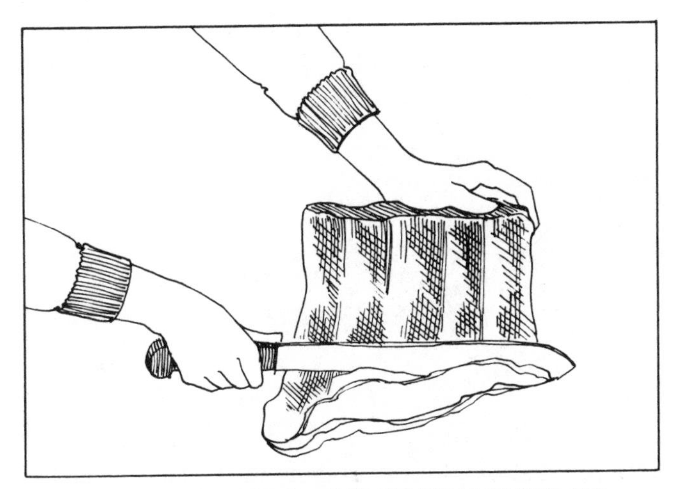

B-22. STANDING PIECE ON END TO TRIM OUTER FAT

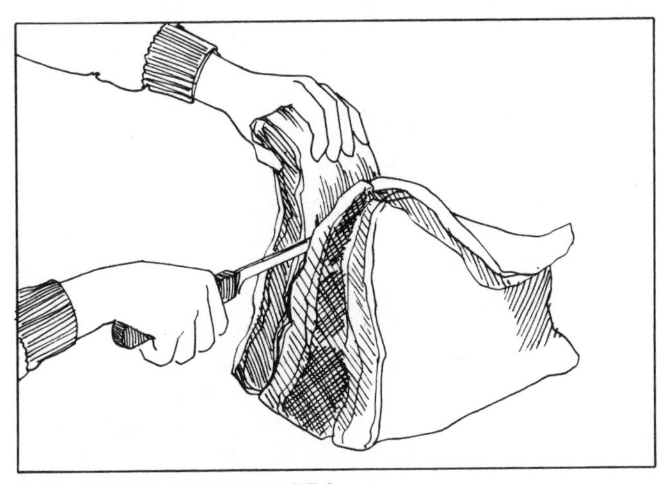

B-23. SAWING INTO THIRDS

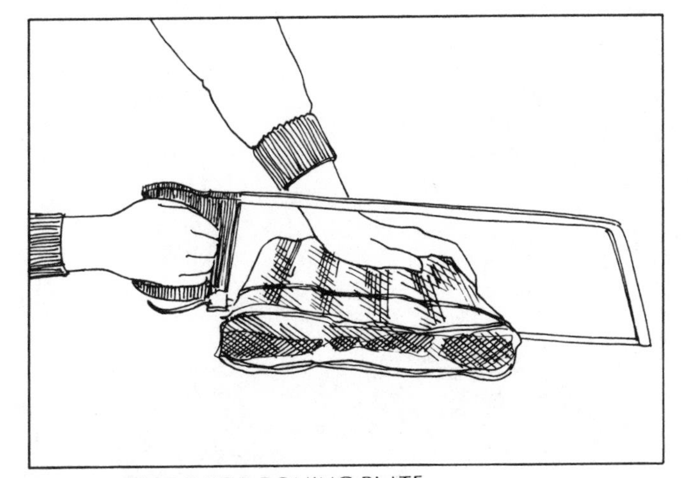

B-24. METHOD FOR BONING PLATE

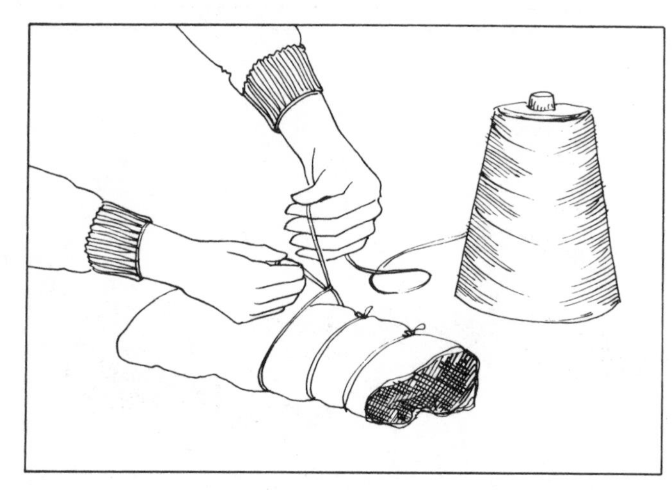

B-25. TYING FOR POT ROAST OR CORNED BEEF

remove all the fat in and around the meat. The finished piece will be the boneless plate. Pastrami is made from this piece.

Fold the piece in half, then tie so that one half folds over the other half to make a double thickness. Tie in at least 6 places down its length. The piece is now ready for pot roast or for corned beef (B–25).

PLATE BONES

Using the saw, after trimming fat off all bones, cut in thirds and use for making stock (B–26).

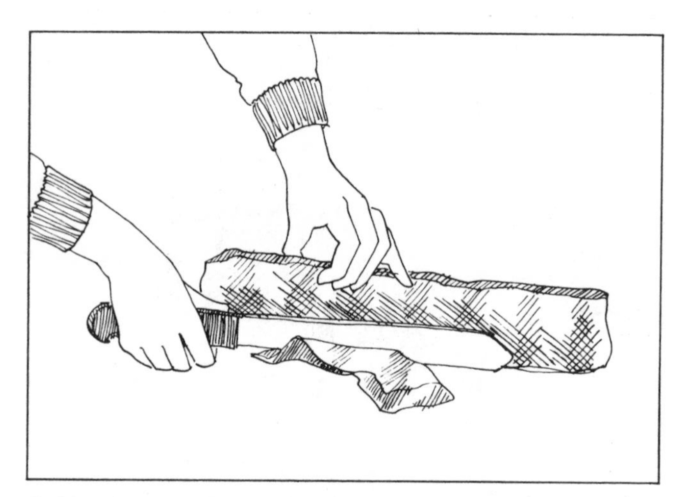

B-26. TRIMMING AND CUTTING FOR STOCK

CROSS-CUT CHUCK

This is the other part of the forequarter you first cut into two sections: the rib and plate, already discussed, and cross-cut chuck with brisket and shoulder. Now cut the remaining large piece again into two sections—the chuck and the arm and shoulder. (Once again, you can buy these sections already cut and proceed from there following the instructions detailed below.)

With one hand, lift the arm (foreleg) up and down to feel the movement in it. Press your other thumb into the very middle of the entire cross-cut chuck as you lift the arm up and down (B–27). You'll feel the socket joint wiggle. Make a small cut, a mark for yourself, right against where you feel the end of this joint.

Go to the other side of the whole piece. There you will see a white cartilege running horizontally. Measure approximately 1 inch from the end of this cartilege and make a second mark with your knife (B–28).

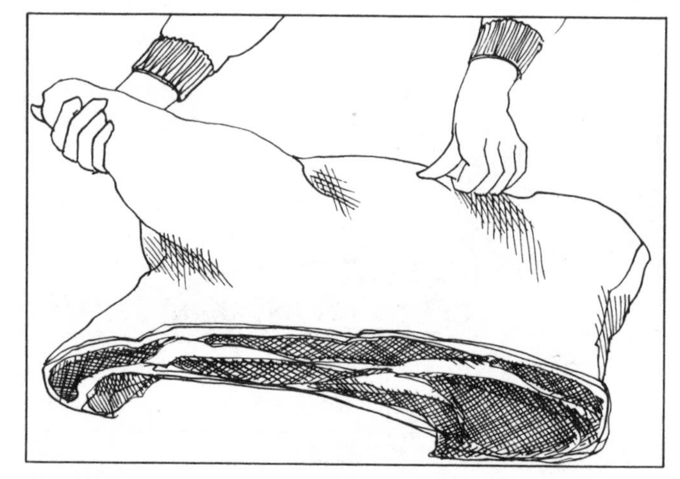

B-27. FINDING THE SOCKET JOINT

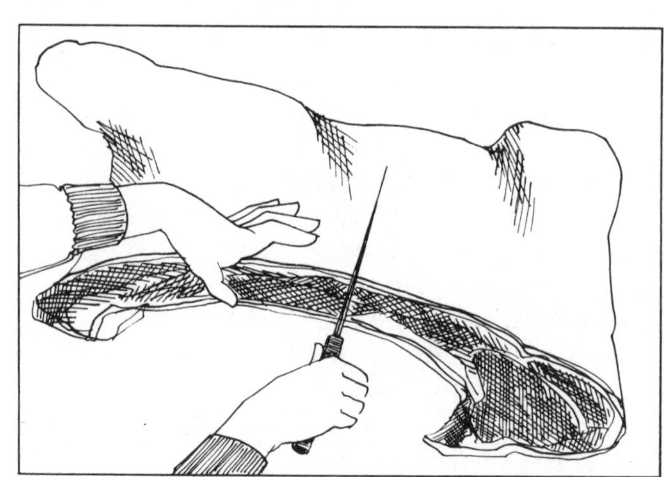

B-28. MAKING MARK OPPOSITE JOINT

Observe the first mark you made on the piece and line up both. They will be your guideline for separating the arm and shoulder section from the chuck section. With the 12-inch knife, cut along this line. Cut down all the way across as far as your knife will go. The blade should sink all the way into the piece until you reach bone. Then with the saw in the same incision, cut through the bone (B–29). Stop when you have sawed through the bone, and with the knife finish cutting completely through the piece into two sections—the chuck and the arm and shoulder. The arm and shoulder section will be larger than the chuck section. Remember, in our way of cutting this piece, the brisket and shank are part of the arm and shoulder.

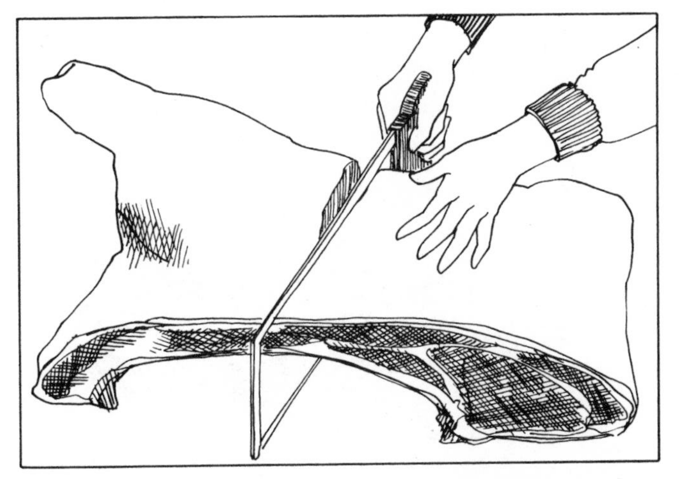

B–29. SAWING THROUGH THE BONE, USING THE TWO MARKS AS GUIDES

ARM AND SHOULDER

From this section, you will cut meat for:

Brisket (1st and 2nd cuts)
Breast Flanken
Shoulder Roast (Arm Roast)
Arm Steaks (Shoulder Steaks)
Boneless Shoulder Roast
Boneless Shoulder Steaks (Boneless Arm Steaks)
Shin Beef
Soup Meat
Marrow Bones

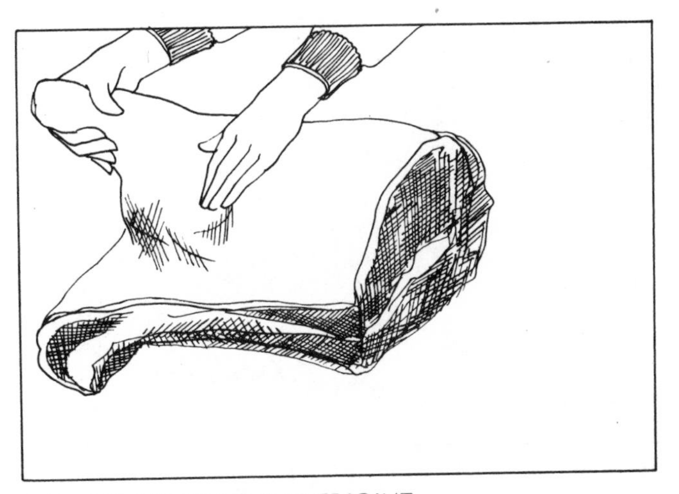

B–30. FEELING FOR THE KNEE JOINT

Separate the arm and shoulder from the brisket and breast flanken. Lay the piece down with the shoulder (foreleg) on the top side. Move the leg and feel for the knee joint (B–30). You will see two horizontal portions of meat. Place your fingers at the corner diagonally across from the foreleg and begin to pull back while cutting with the knife into the fat and membrane (B–31). But cut only to where you can see the beginning of the contours of the foreleg appear inside the cut. Turn the blade to follow the natural contours of the foreleg, cutting into the fat while with your other hand you pull back on the thick, heavy foreleg (B–32). Continue to cut, going in deeply, creating a deep cavity around the edge of the foreleg (B–33). Keep cutting and pulling and soon the entire foreleg portion will come free. Lay this back and the whole piece will be laid open like a giant butterfly (B–34). Keep cutting, making small cuts, as you continue to separate the 2 sections until you have completely separated the arm and shoulder

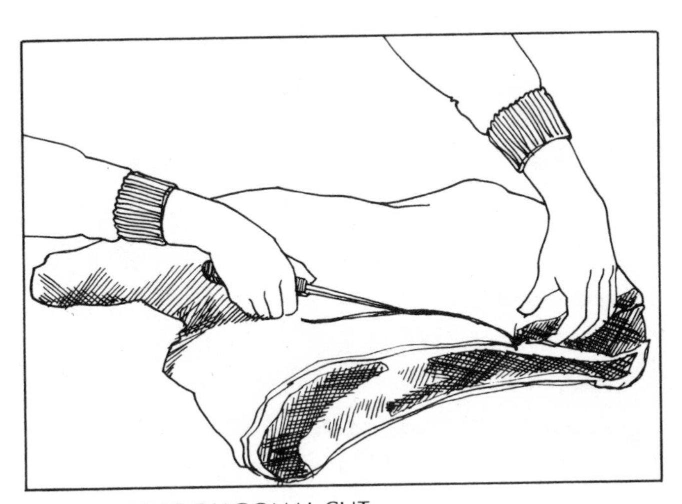

B–31. MAKING DIAGONAL CUT

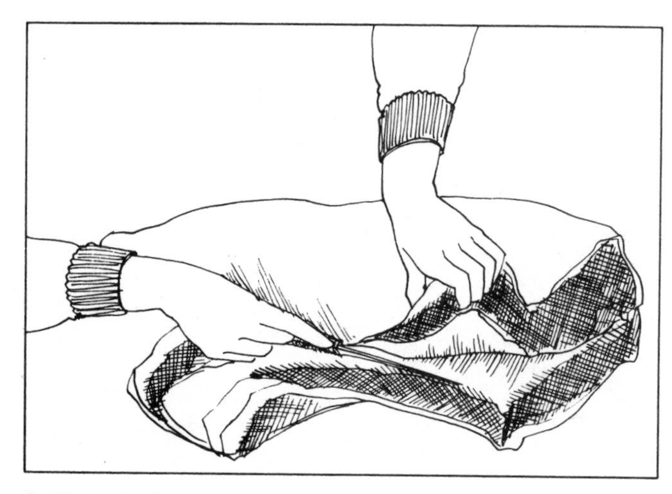

B-32. WORKING MEAT BACK TO TAKE OFF SHOULDER

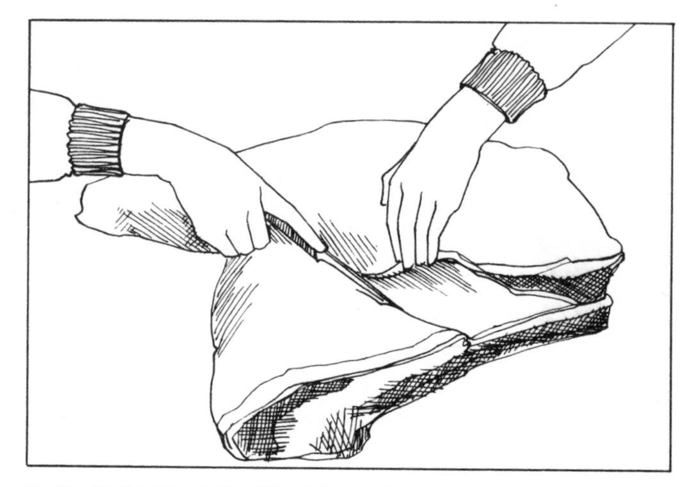

B-33. FOLLOWING CONTOUR OF FORELEG

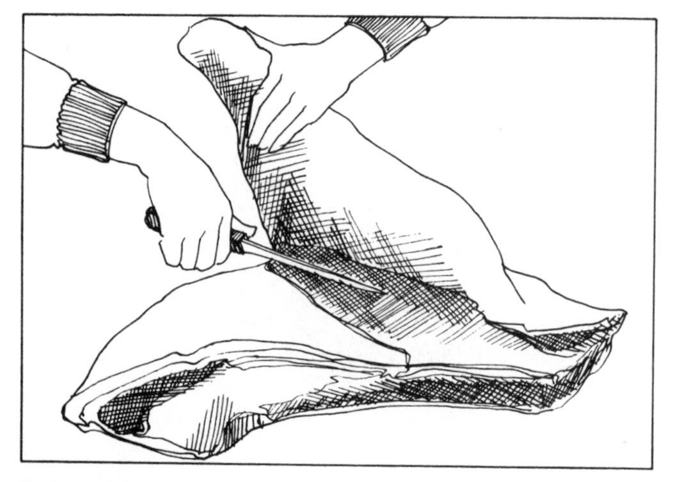

B-34. LIFTING AND SEPARATING FORELEG SECTION FROM BRISKET

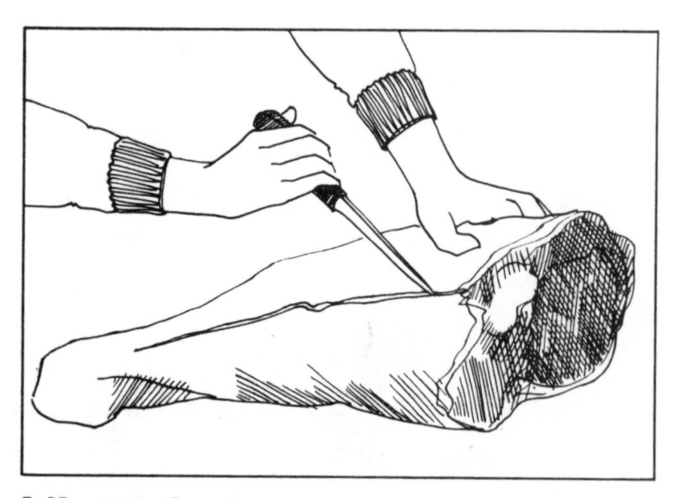

B-35. MAKING INITIAL CUT TO SHOULDER BONE

(foreleg) section from the brisket. (Remember, you can purchase these parts already sectioned at your bulk market, but it is far more economical to section them yourself.)

Set aside the brisket and breast flanken piece.

SHOULDER

The first step is to bone out this large shoulder piece. Lay the shoulder on your cutting surface with the flat, horizontal piece of bone facing you. Start at the top of the bone nearest you and begin to cut in (B–35). Continue cutting in until you see the end of the large shoulder bone. Using the end of the shank, turn the piece completely around and start cutting in deeply along the line of the shoulder (foreleg) bone. As you do you will open up the side of the meat but you'll find you cannot cut all the way through because you must go around the bone.

Return to the top of the shoulder where you first started to cut. But now you will cut around the other side of the large, bulbous shoulder bone (B–36). Follow the line of the bone, pulling back the meat with your other hand as you cut. Soon you'll reach the lower part of the bone, past the elbow joint. Now you should have the meat of both sides of this shoulder (foreleg) bone cut away and the bone exposed. You will now be near or at the end of the shank bone.

Return to the shiny ball and socket. Brace the end of the shank against your leg to help you lift the entire piece a little as you continue cutting (B–37), now *under* the entire length of the bone. Cut as close to the bone as possible (B–38), leaving as little meat on the bone as you can. Remember to work around the elbow, using the tip of your knife. Finish by severing the very end of the bone.

Now take off the long, flat portion of meat you have cut from around the bone, separating this from the thick piece (B–39). Set aside. Cut out the small triangular bone remaining in the thick part (B–40).

The thick piece of meat that is left is the boneless or shoulder steak. It is also called the cross-rib section, though there are no ribs in it. You will see a large, fatty piece at one end. Remove this by seaming it out (B–41).

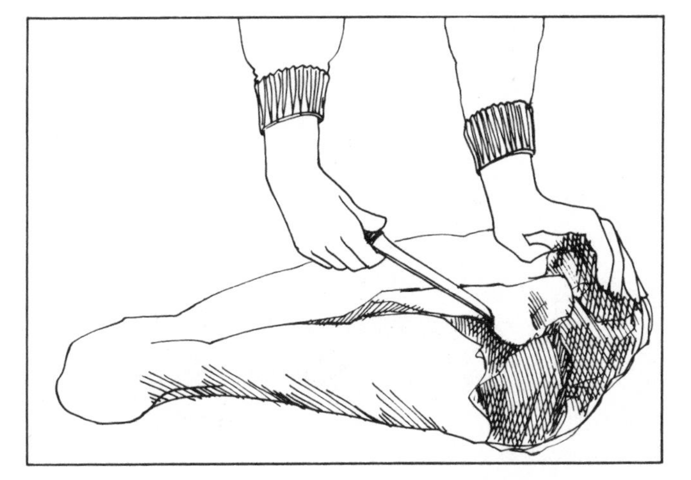

B-36. LOOSENING BULBOUS SHOULDER BONE

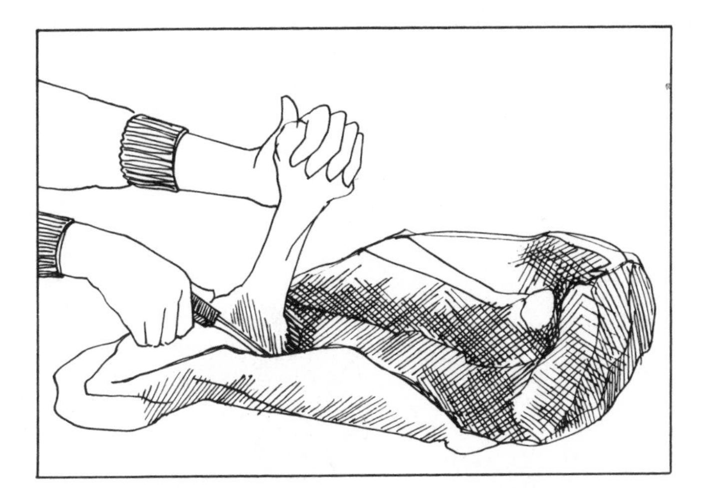

B-37. WORKING AROUND BOTTOM OF BONE

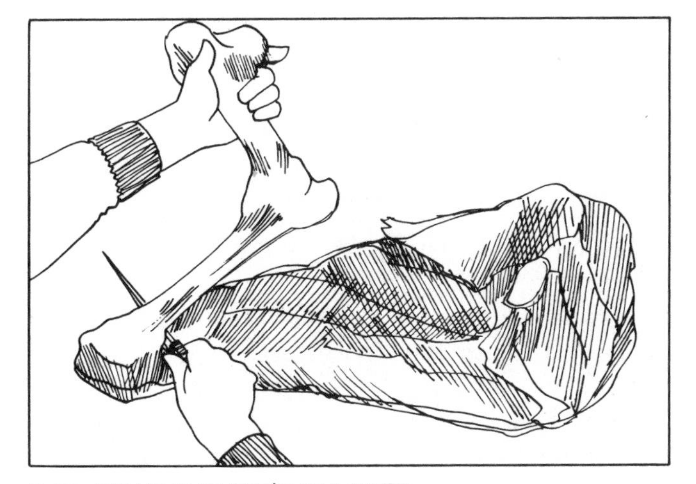

B-38. REMOVING BONE IN 1 PIECE

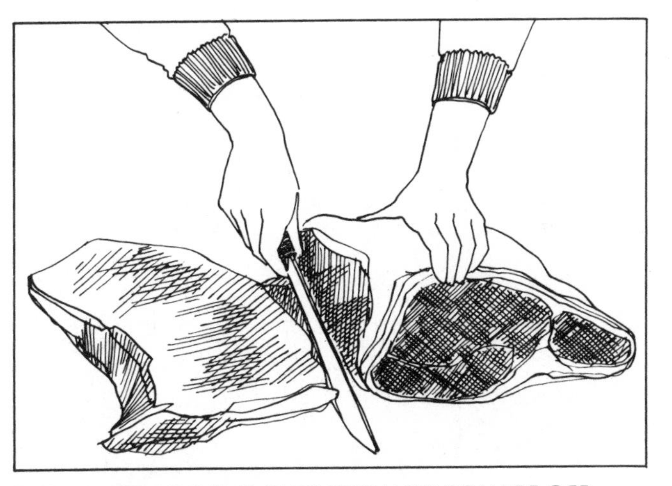

B-39. CUTTING OFF FLAT PORTION TO SQUARE OFF SHOULDER

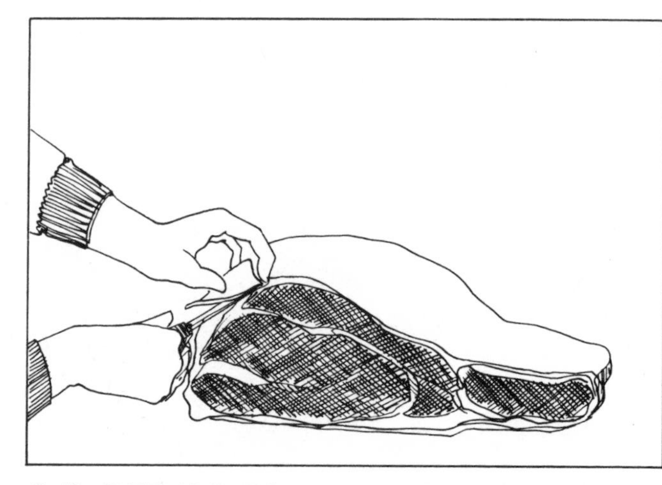

B-40. CUTTING OUT SMALL TRIANGULAR BONE

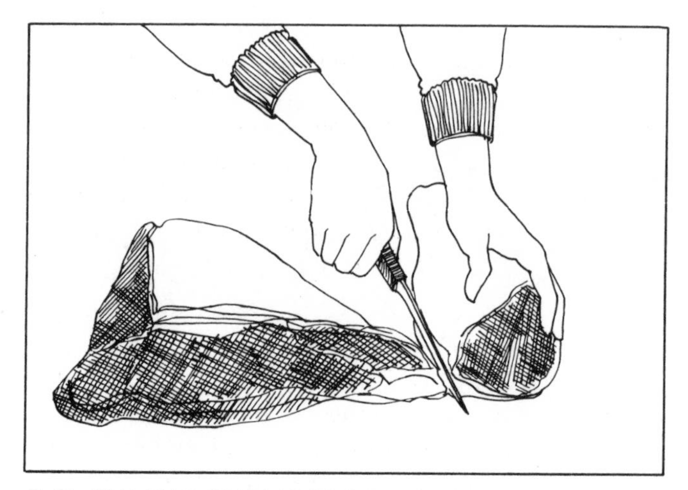

B-41. SEAMING OUT LARGE FATTY PIECE

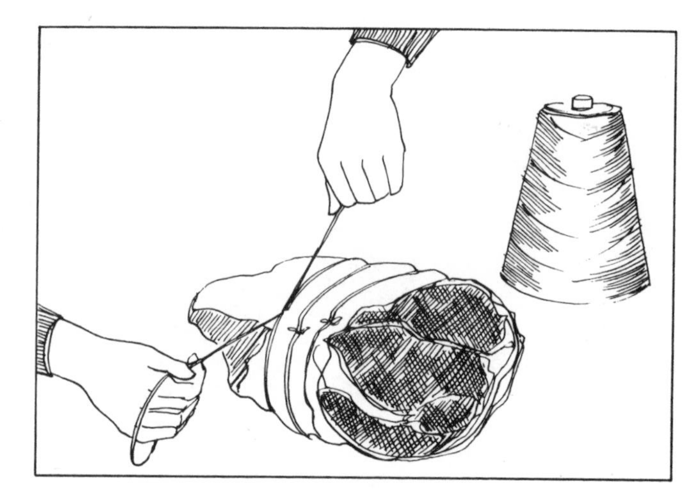

B-42. TYING BONELESS SHOULDER

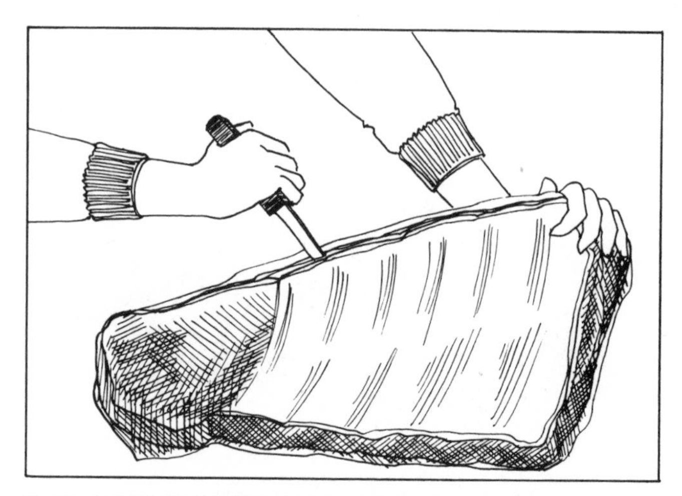

B-43. MAKING INCISION AT TOP OF BREASTBONES

BONELESS SHOULDER ROAST

This is the large, thick piece of meat that is left. It will be easier to handle this piece if you tie it now. Make at least 10 ties (B–42). When finished, the boneless shoulder roast is ready for the oven.

SHOULDER STEAK

For boneless shoulder steaks, just slice down in the desired thickness. If the roast has been tied, the steaks will have a nicer shape.

HAMBURGER

All leftover pieces of meat may be ground for hamburger. *Note:* If mixing in meat from the very end of the shank piece you must remove the tough, outer skin from around the meat of this area.

BRISKET

The brisket is a large piece of meat that is thin on one side and curves upward into a thick area. Place flat with the rib side up. Your first step is to bone out the piece, removing the breastbones.

Make an incision with the knife along the line of the top of the breastbones, cutting in some 2 inches through the fat until you see the meat below, all the while pulling the incision wider with your other hand (B–43).

Now turn the piece around so that the curve of the breastbone faces you. Cut deep and under to cut away the entire breastbone section (B–44). When you've cut under far enough, you can pull it up with your hand, just cutting lightly across the thin connecting tissue until the complete breastbone section comes off (B–45). What is left is the brisket.

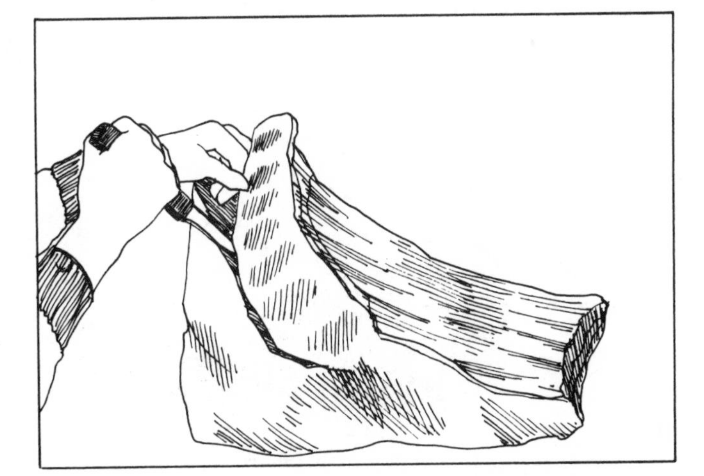

B–44. CUTTING ON OPPOSITE END, USING HAND TO WIDEN INCISION

BRISKET

Place the brisket on the cutting surface. You will see a flaplike piece jutting up out of its center. Trim all excess outer fat from around this flap. The brisket will have a thinner and a thicker portion. The thinner portion is the first-cut brisket; the thicker part is the second-cut brisket. The top layer of the second cut has more fat in it.

Cutting the brisket exactly in half at an angle gives you a little of the second cut in the first cut (B–46).

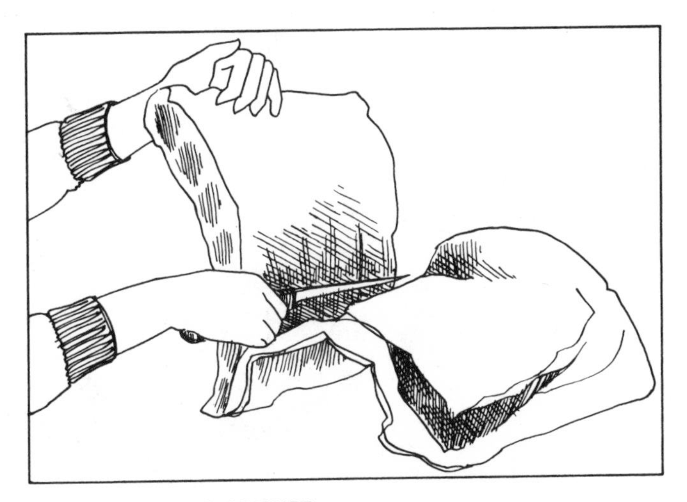

B–45. REMOVING BRISKET

BREAST FLANKEN

This is divided into 2 sections. Saw through the breastbone at an angle (B–47) so that the two separated pieces will both be somewhat triangular. Trim both pieces.

The meatier portion of the 2 pieces is to be sawed again after trimming (B–48). This time saw the center the long way, but not completely through. When you've sawed through the bone, cut, now with the knife, *across* the lengthwise incision made with the saw (B–49). These will be the breast flanken pieces, for use in soups and slow barbecue dishes.

The upper breastbone piece is still left. Trim away any remaining fat, then saw and cut into 4 sections and use for beef stock (B–50).

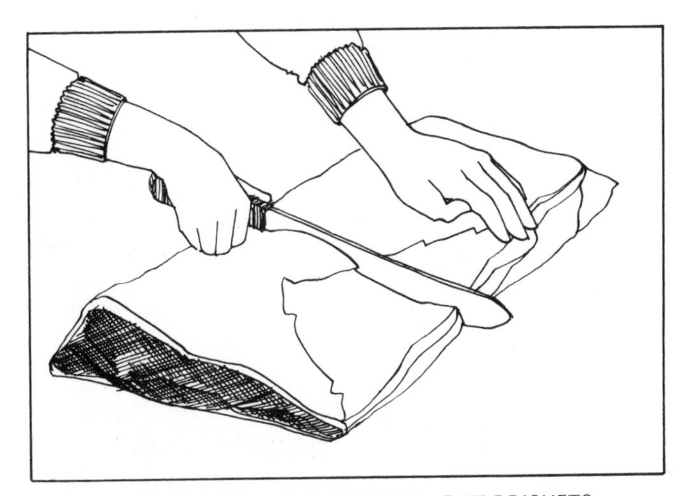

B–46. MAKING FIRST- AND SECOND-CUT BRISKETS

B-47. SAWING THROUGH BREAST FLANKEN

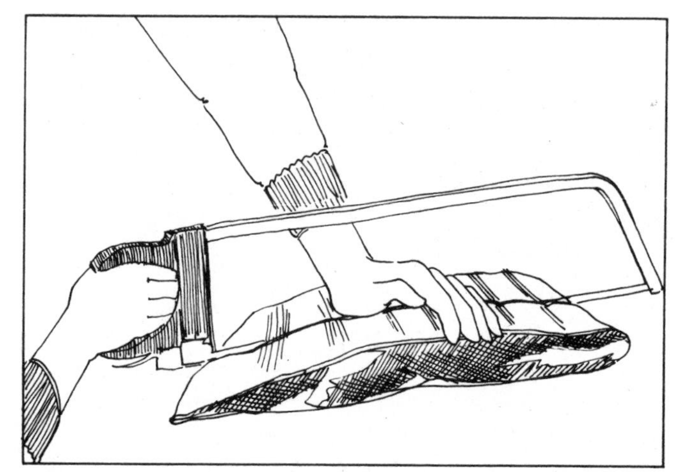

B-48. SAWING MEATIER PORTION IN HALF

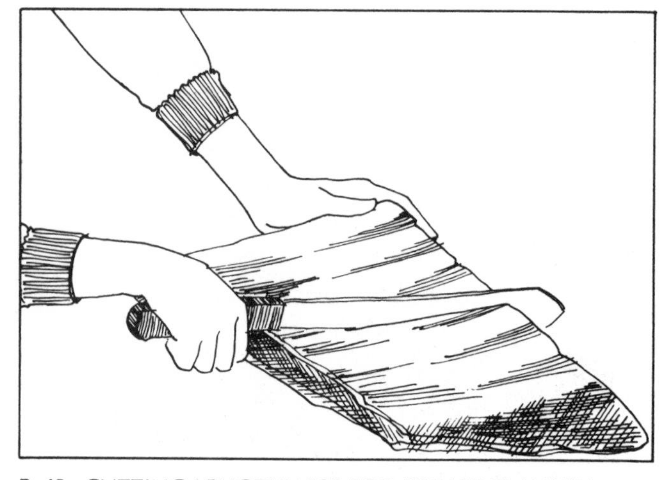

B-49. CUTTING LENGTHWISE FOR BREAST-FLANKEN PIECES

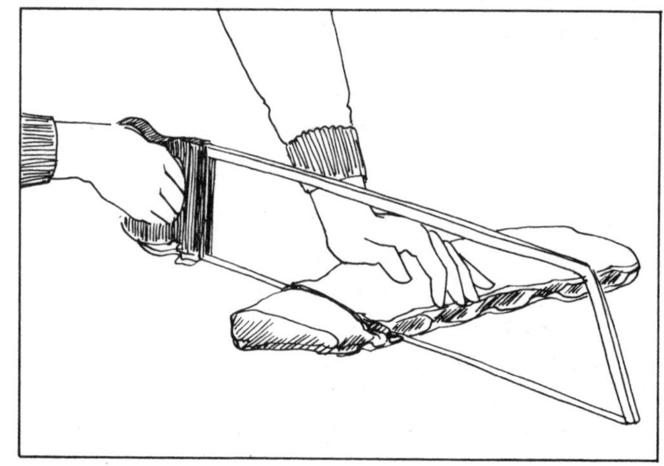

B-50. MAKING PIECES FOR STOCK FROM UPPER BREASTBONE PIECE

CHUCK

From this large piece you will cut meat for:

Chuck Roast
Chuck Steak
Chuck Pot Roast (Bone In)
Boneless Chuck Pot Roast
Chuck Stew
Ground Chuck

In the way we have sectioned everything up to now, this large piece can be simply and easily cut into the pieces for the dishes listed above. Some cuts of chuck often seen in retail markets have been eliminated because we feel that our method provides the best use of the chuck.

CHUCK ROAST

Place fat side up on the cutting surface, the meat side away from you. Look at the meaty side and measure off 2 rib bones. Place the 12-inch knife between the 2nd and 3rd bones and cut straight down to the bone (B–51).

Finish the incision with the saw to cut the bones. When you no longer feel the bone against the saw, stop at once, take out the saw and return to the large knife. Spread the cut apart with thumb and forefinger and cut with the knife until you reach more bone. Then take the saw once more and finish cutting through the bone. Use the knife once more to finish the complete incision. The reason for the sequence of knife and saw, knife and saw, is to avoid tearing the meat with the saw.

Lift off the top section you've cut (B–52), severing it completely. Set aside for stew and ground meat.

The entire remaining piece is the chuck roast. You can use it as is for a large gathering or separate this piece into smaller roasts. Using the 8-inch knife, cut in at the end just above the bones and cut and pull off similar to the way you removed the top section. When finished there will be 2 roasts (B–53).

We suggest leaving the bone in for this piece because it adds flavor. Also, if you are freezing it, keeping the bone in helps retain the juices.

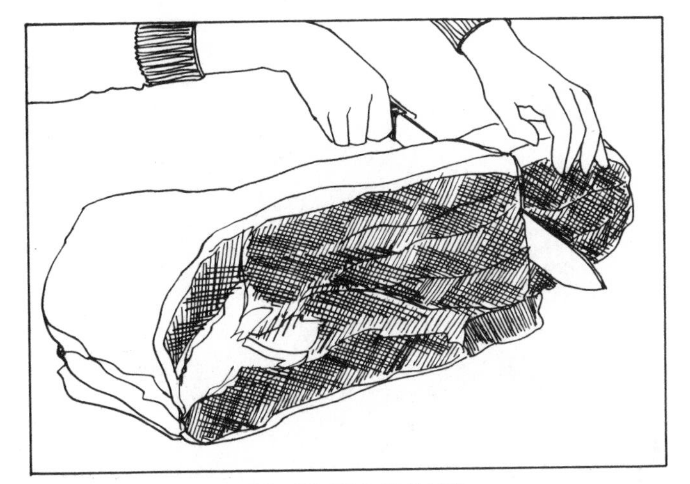

B-51. CUTTING RIB BONES FOR CHUCK

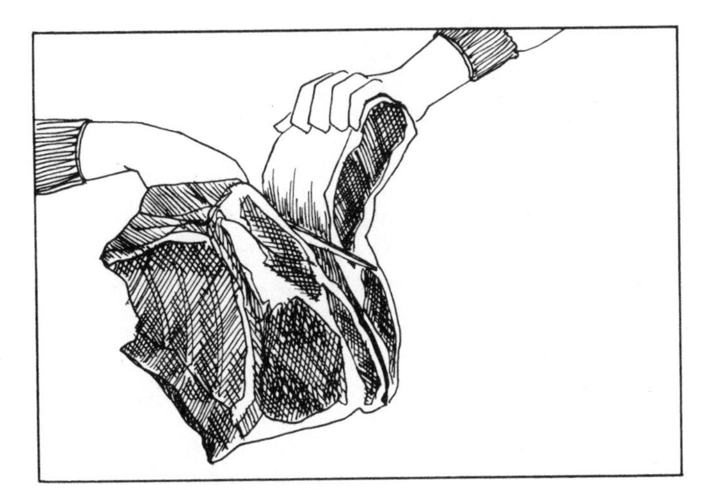

B-52. USING KNIFE AND HAND TO SEPARATE AND LIFT OFF TOP PIECE

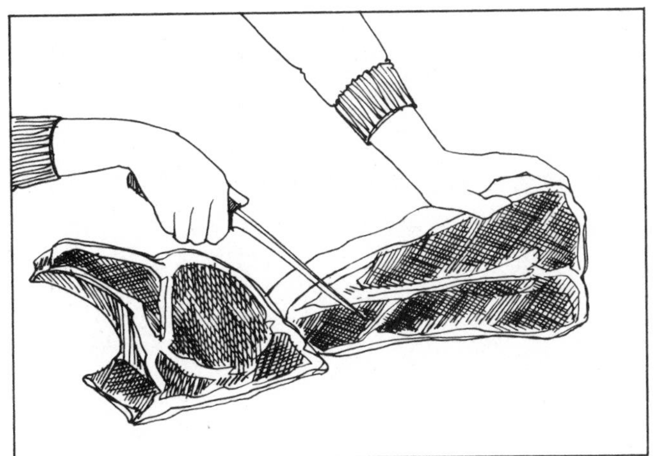

B-53. MAKING 2 ROASTS OUT OF CHUCK PIECE

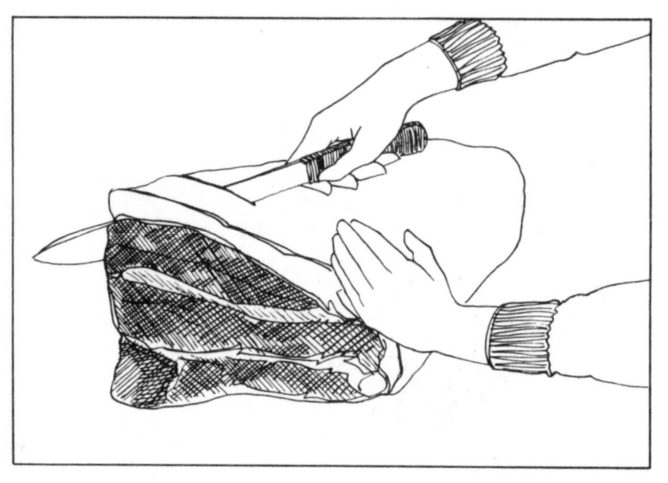

B-54. CUTTING CHUCK STEAKS

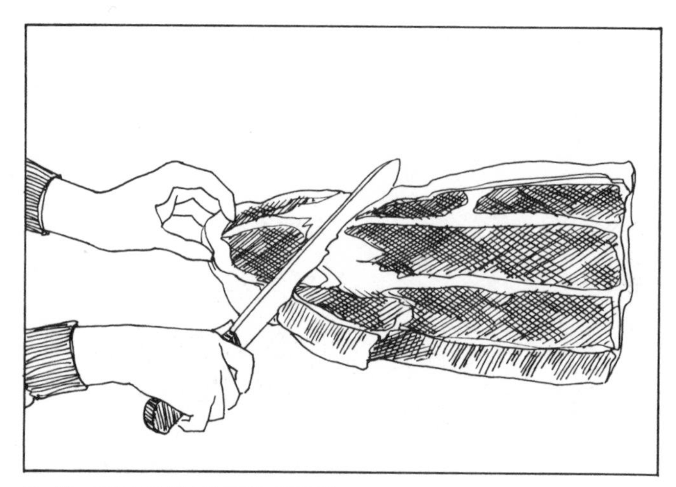

B-55. REMOVING TAIL PIECE

CHUCK STEAK

For chuck steaks, use the 12-inch knife to cut down from the end of the piece, cutting from the top to the desired thickness. You should be able to cut four 1¼-inch-thick chuck steaks from the main piece. Again, remember to alternate knife and saw to protect the meat—saw for bones, knife for meat (B–54).

Take off the tail end and set aside for stew and ground meat (B–55).

CHUCK POT ROAST (BONE IN)

Lay the chuck fat side up with the meat facing away from you. You will see a piece of flat bone that resembles an inverted T. Place your finger under this bone and lift and you'll feel this section come up. Separate by cutting into this seam or separation line as you pull back with the fingers of your other hand (B–56). Continue cutting into the thin tissue, and pull and cut, pull and cut, until the entire piece is almost free (B–57). Sever the last piece of fat holding it to the main section. You'll have a thick triangular piece with the inverted T-bone attached. This is the chuck pot roast bone in.

Trim the entire outer layer of fat from the rest of the chuck section still attached to the main section (B–58). Now turn the entire section of chuck over, so the underside bone part is facing you. With the saw, cut into the bone at an angle, to make 3 or 4 equal sections. Use the other hand to grip the bone for support (B–59).

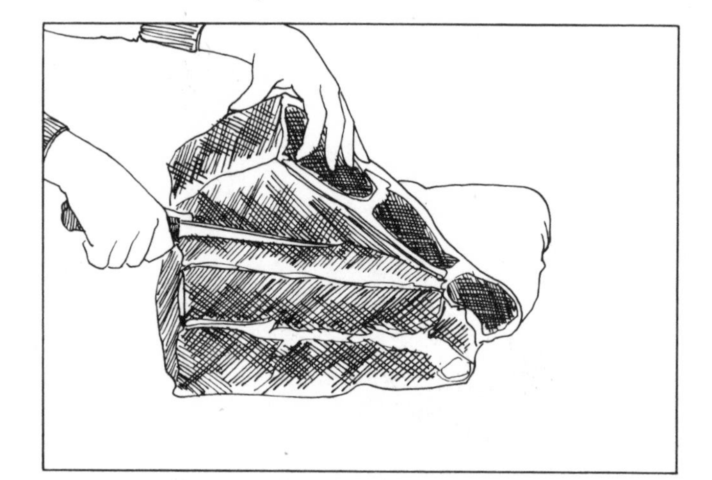

B-56. USING HAND AND TIP OF KNIFE AT NATURAL SEAM

B-57. TAKING OFF BONE IN CHUCK POT ROAST

B-58. TRIMMING FAT FROM MAIN CHUCK SECTION

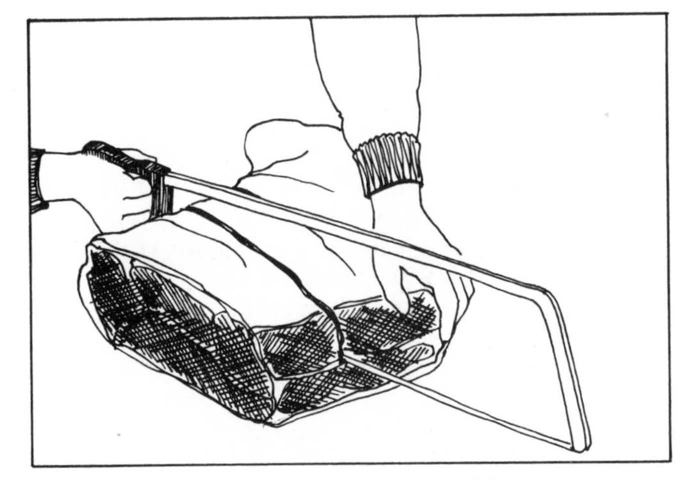

B-59. SAWING PIECES FOR CHUCK ROAST

BONELESS CHUCK POT ROAST

One of the pieces you have cut is the neck piece. Set this aside. Bone each of the remaining pieces the same way: Stand the piece, bone down, on the table. Go to the end of the piece and insert your 8-inch knife *flat* over the bone. Cut in along the top of the bone, following the contours of the bone until you have cut off the entire bone (B–60). Save the bones for stock. Trim the other end of the piece and cut the deep-yellow spinal cord away (B–61). Tie the boned piece with 3 ties, 2 across the width and 1 lengthwise (B–62).

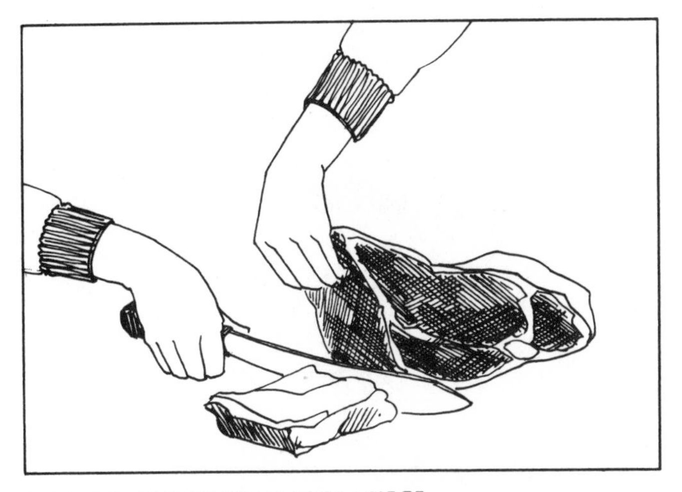

B-60. REMOVING BONE FROM PIECE

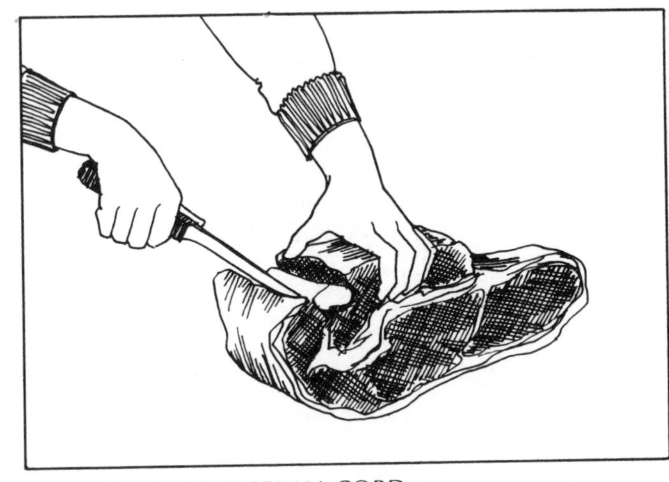

B-61. BONING OUT SPINAL CORD

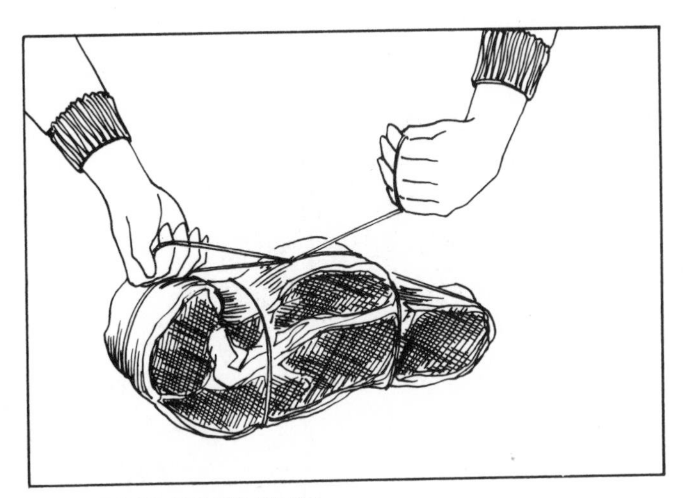

B-62. TYING BONED PIECE

NECK

With the saw, cut into 2 or 3 equal pieces (B–63, B–64, and B–65). Then, with the knife, remove the bones, following the natural line of the bones. Grind the boned pieces for excellent, sweet hamburger. With the 12-inch knife, cut one of the boned neck sections into 2-inch pieces for stew meat (B–66).

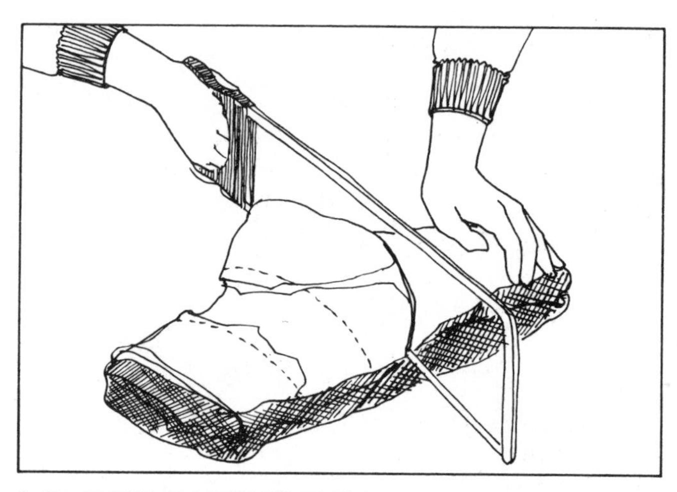

B–63. CUTTING NECK PIECE FROM REST OF CHUCK

B–64. CUTTING NECK PIECE IN HALF

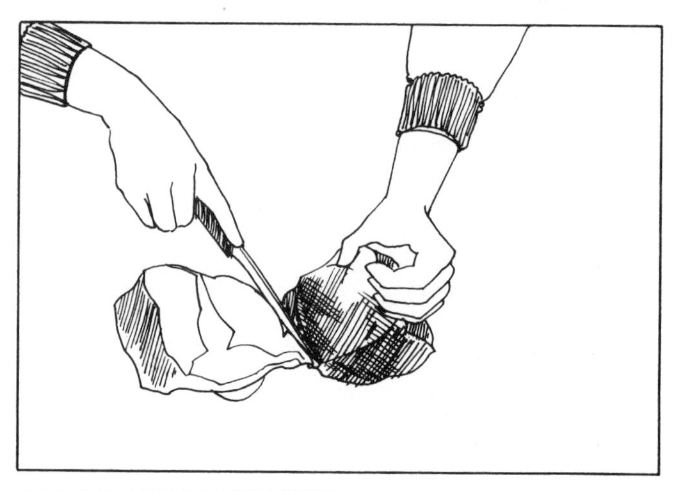

B–65. TRIMMING NECK PIECE

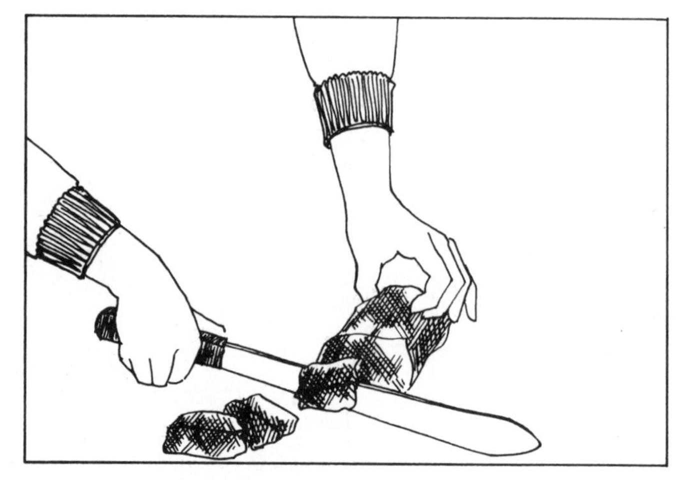

B–66. CUTTING CUBES FOR STEW

SHORT LOIN

The hindquarter is made up of 3 pieces. These can be purchased as (A) the entire hindquarter; (B) the Short Loin section; (C) the Sirloin section and hip; (D) the Round (B–67). To use any of these sections, follow the instructions given here for that specific section.

From the Short Loin you get:

Delmonico Steak (1st cut)
T-Bone Steak
Porterhouse Steak
Short Fillet Roast
Chateaubriand
Filet Mignon
Fondue Pieces
Shell Steak
Full Fillet
Fillet Roast
Boneless Shell Roast
Boneless Shell Steak
Tail of Shell

Place the short loin with outer fat down (top side down). The chine bone will face you. Of the 2 ends, one will be longer and have the tail of the meat on it. Face this away from you. Trim the inner fat from the entire short loin, which encloses the kidney (B–68). You may saw the tail section off before cutting steaks (B–69).

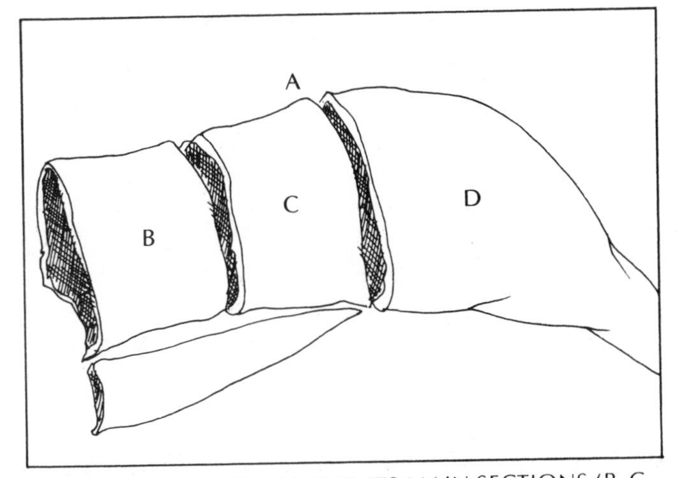

B–67. HINDQUARTER (A) AND ITS MAIN SECTIONS (B, C, AND D)

B–68. REMOVING KIDNEY FAT

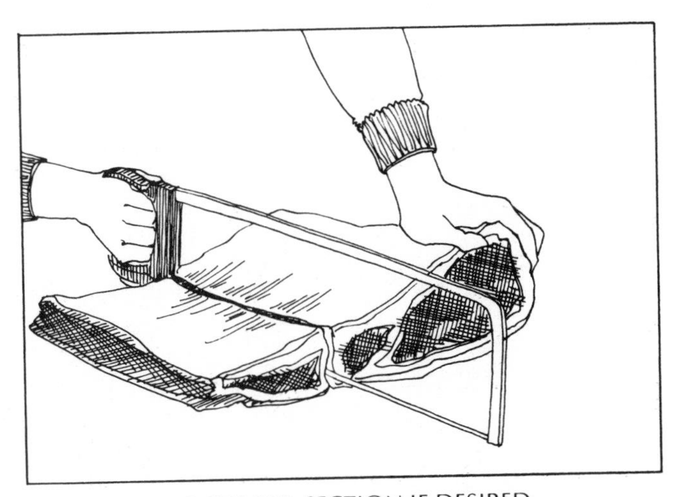

B–69. CUTTING OFF TAIL SECTION IF DESIRED

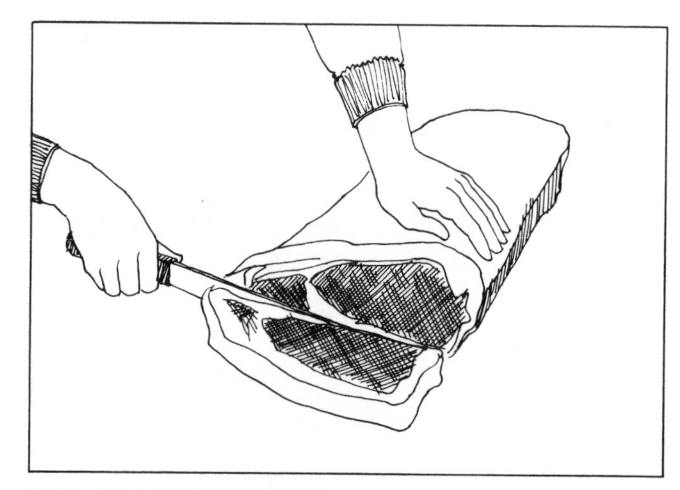

B–70. DELMONICO STEAKS

DELMONICO

Cut from the eye, the Delmonico section, the smaller end without the fillet showing, is about 3 inches from the end. To remove this section, use the knife and saw (B–70). When you have cut it off, trim the fat away. You should be able to cut three 1-inch Delmonico steaks from this section.

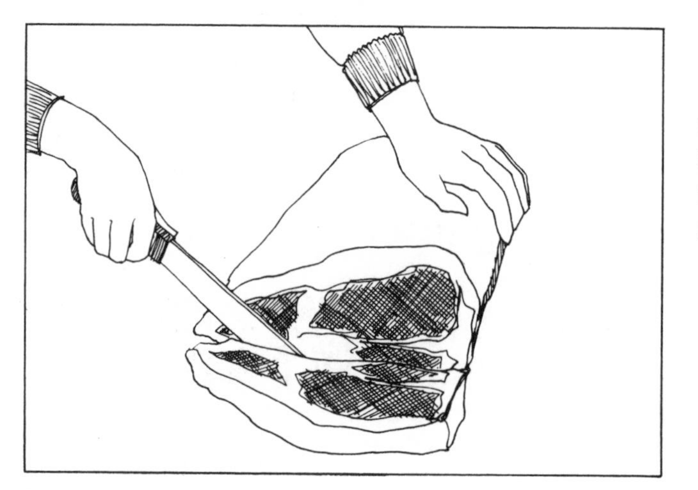

B–71. T-BONE STEAKS

T-BONE STEAK

After the Delmonico section is removed, you come to the T-bone section and should see the T-bone itself. Cut to the desired thickness through entire section for T-bone steaks (B–71). This T-bone section is about 4 inches long and makes four 1-inch steaks.

PORTERHOUSE STEAK

The section left now is the porterhouse. Of the 3 sections, the porterhouse section has the largest amount of the fillet. Cut the porterhouse about 1¼ to 1½ inches thick (B–72). You should get 4 thick porterhouse steaks from this section. After steaks are cut, trim the fat but leave a little on and score it. (Score the fat only, not the meat.) Scoring allows you to curl the tail neatly around the end of the porterhouse and the fat to crisp evenly (B–73).

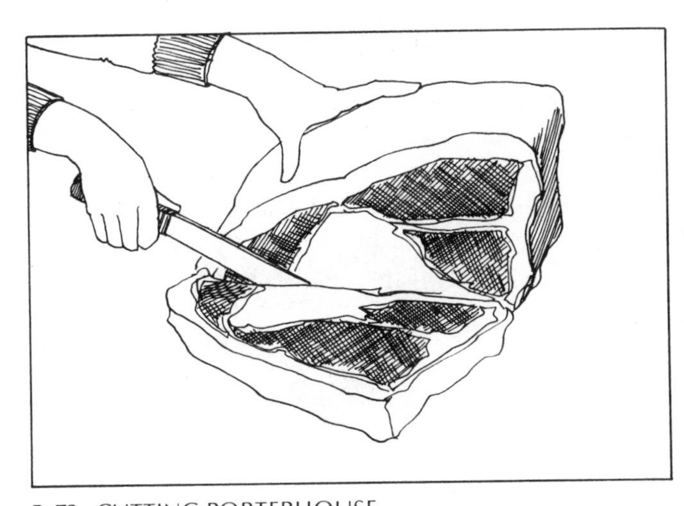

B-72. CUTTING PORTERHOUSE

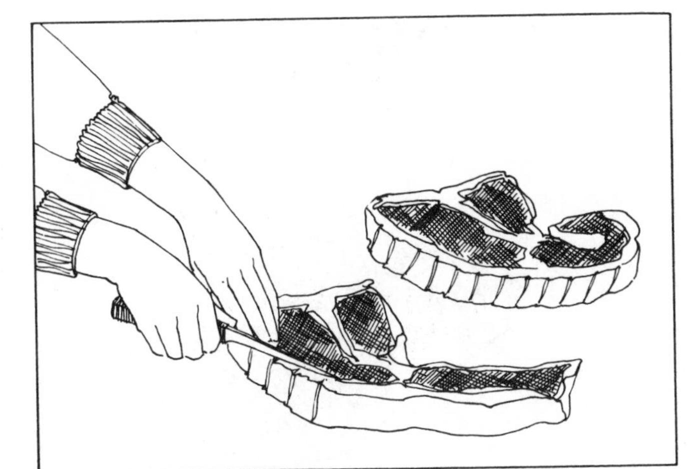

B-73. SCORING FAT ON STEAKS

FILET MIGNON

For the filet mignon, return to the entire short loin piece. From this entire piece you will be able to cut your short fillet loin. First trim the inside cavity of all excess fat. Then trim off the top of the backbone so you can see the backbone clearly (B–74).

Removing the fillet: Hold your other hand on the eye of the fillet, starting at the end farthest away from you. Weave the top of the knife blade in and out along the wavy top of the backbone segments (B–75). You'll have to keep digging in with the top of the blade to run along this wavy backbone, and you will need at least 3 runs down the entire length of the backbone to completely loosen it.

After you've made a deep incision along the full length of the backbone, place the knife at the end nearest you. Work around and on top of the bone, pulling the meat back away from the backbone (B–76). When you've gone about 6 inches, reverse the entire piece and put 4 fingers into the incision you've made. Press with your fingers against the backbone while pushing with your thumb against the meat (B–77). Insert the knife and, working from the other end, go under and around the entire fillet.

When you have cut the fillet out completely, fat is still attached. Trim off all the outer fat. You are left with a whole short fillet, the thick part at one end, the rest tapering down on the other end.

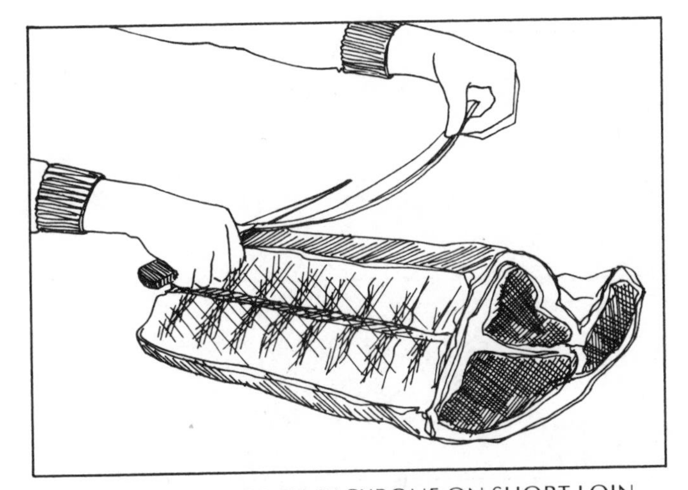

B-74. TRIMMING TOP OF BACKBONE ON SHORT LOIN

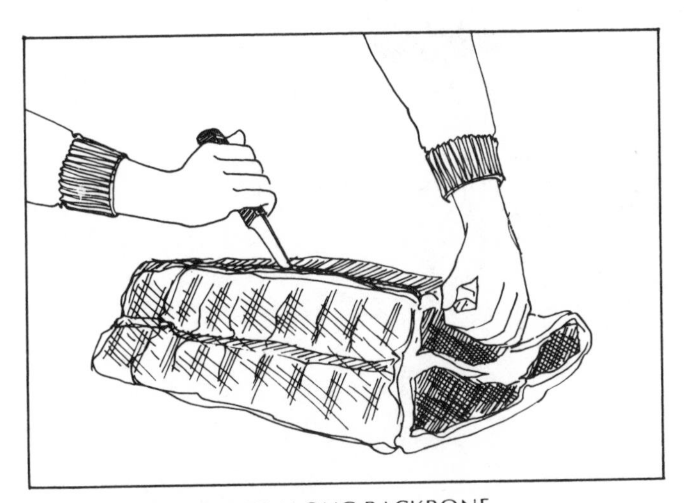

B-75. WEAVING KNIFE ALONG BACKBONE

B-76. USING KNIFE AND HAND TO PULL AWAY FILLET

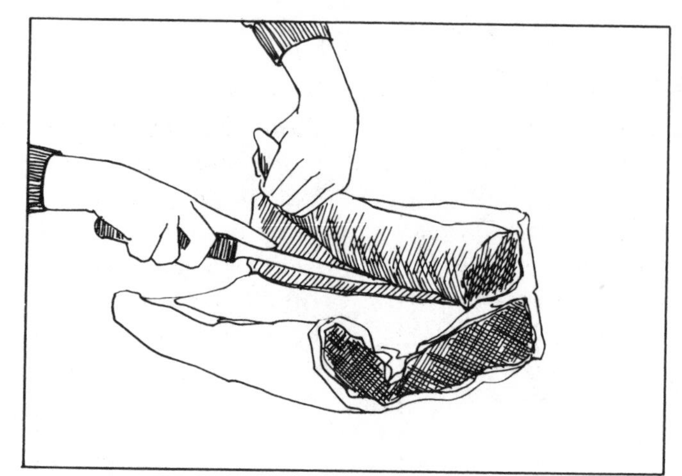

B-77. PULLING FILLET UP TO REMOVE COMPLETELY

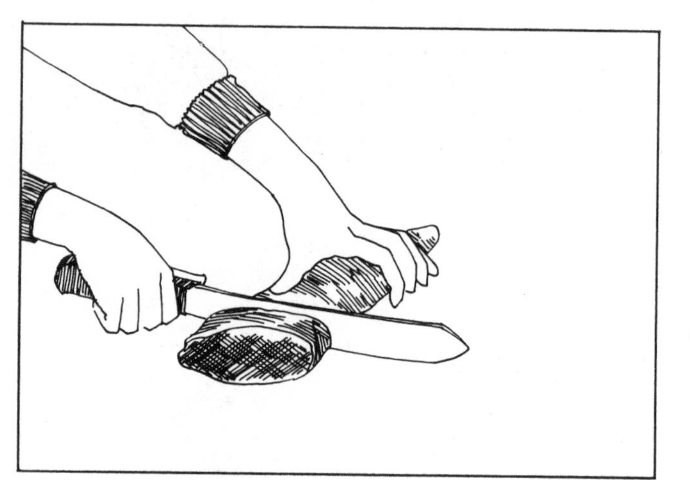

B-78. CUTTING FILLET AT THICK END FOR CHATEAUBRIAND

SHORT FILLET ROAST

For this, use the entire piece after you've properly trimmed it.

CHATEAUBRIAND

Cut the thick half of the short fillet section in two for Chateaubriand (B-78).

FILET MIGNON

Slice the short fillet as thin as preferred. You should be able to cut 5 or 6 filet mignons starting at the thick end of the piece.

FONDUE PIECES

As the fillet narrows to the tail, use the remaining part of the tail for small fondue pieces, cubed in ½-inch squares. These may be frozen until enough fondue pieces from other cuts have been accumulated for use in a dish.

SHELL STEAK

After the fillet has been removed from the rest of the short loin, cut down the short loin as for the porterhouse to get the shell steaks.

FULL FILLET

The full fillet, also called full loin, comes out of the short loin and the sirloin sections. Removing both these sections from the sirloin and short loin is too complex and difficult an operation to do successfully at home. It is a job for specialists. To attempt it you risk ruining your sirloin section, and it is usually done only in wholesale markets. We suggest, therefore, that you purchase your full fillet in the sealed package it comes in from the market and trim it yourself.

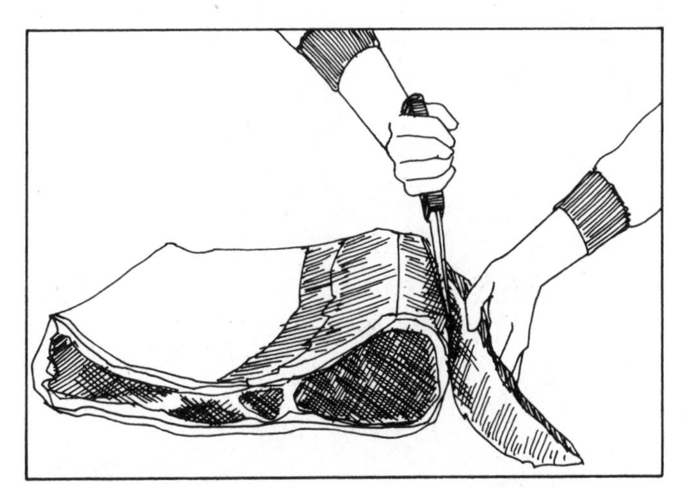

B-79. FINISHING REMOVAL OF CHINE BONE WITH KNIFE

FILLET ROAST

This is the first half of the full fillet. Cut from the entire piece after unsealing and trimming.

BONELESS SHELL ROAST

After removing the fillet from the short loin, proceed to remove the chine bone. First, place the backbone down and position the saw approximately 2 inches above the end of the bones. Saw horizontally, right across the bones (B–79). It will be tough sawing. Do not bear down too forcefully; let your saw do the work.

Saw through this narrow bottom chine bone, and finish the removal of the bone with a knife. Then lay the meat fat side down to cut off the tail. Find the eye of the meat and measure approximately 1 inch from the eye. Saw through the first bone. Make another mark at the other end of the piece, just where you see the meat end, and use the 12-inch knife to take the tail off entirely (B–80). The shell roast will now be a smaller, roundish oblong piece, and some of the backbone pieces will still be along one side. With the 8-inch knife, cut these pieces away. You will also see 6 or 7 little nub bones from the chine bone. Dig these out by using the tip of the blade (B–81).

The flat bones will still be across the entire piece. Keeping the knife flat with the bones, work your way slowly across the piece (B–82), removing the pieces of flat bone as you go. Remember, do not dig deep, but keep your knife flat to retain all the meat you can. You will reach that first rib bone you sawed. Cut down along it and remove. Then, when you've removed the flat bones, trim away all sinews.

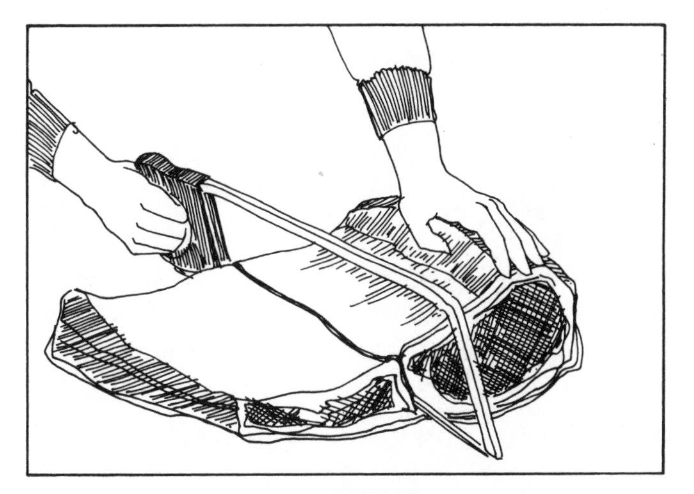

B-80. CUTTING OFF TAIL IF DESIRED

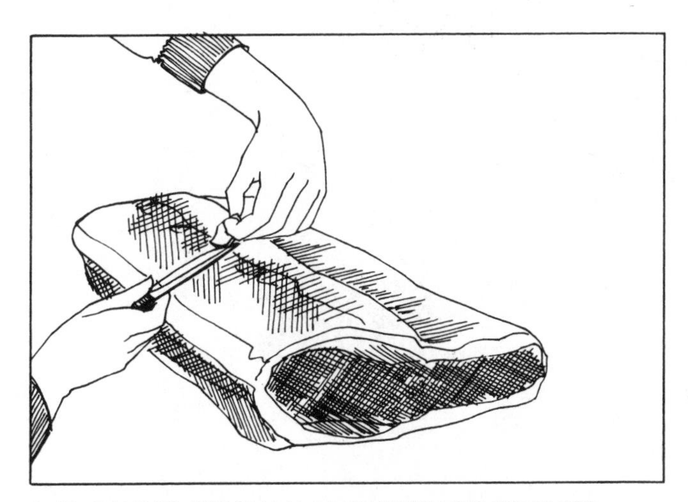

B-81. TAKING OFF SMALL NUBS WITH TIP OF KNIFE

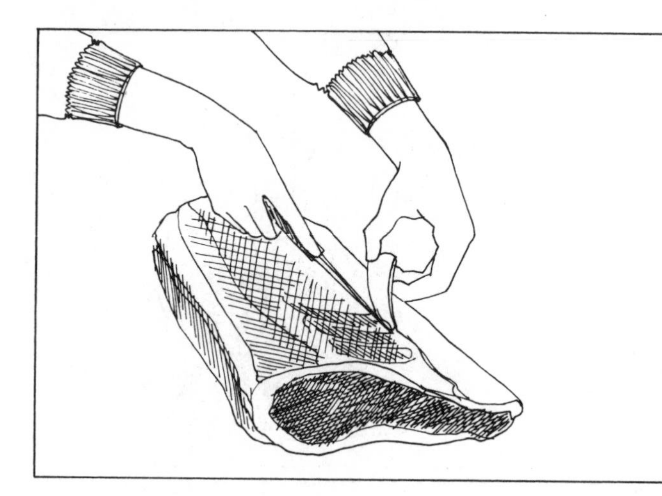

B-82. REMOVING FLAT BONES

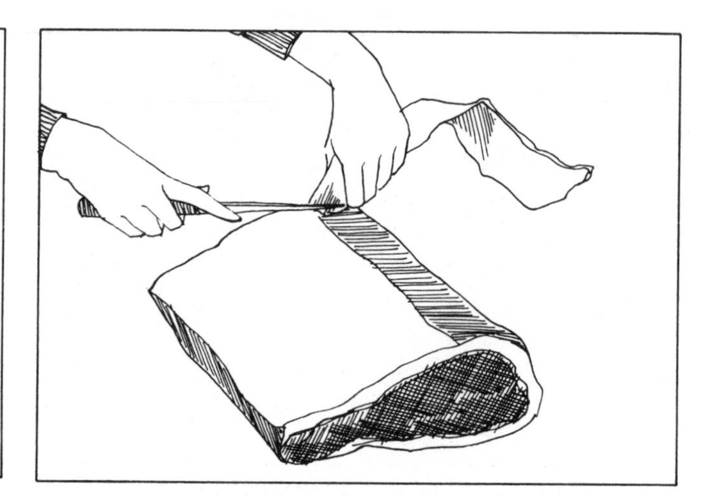

B-83. REMOVING FAT AND VEIN FROM THICK END

Turn the piece over fat side up. Trim off most of the outer fat. At the thick part of the shell roast you'll see a vein. Skin this off along the entire length of the roast (B-83). You must remove all the fat from this side of the top of the shell. Trim some more of the fat from the rest of the shell, leaving about ¼ inch of fat remaining on the shell roast. *Note:* When trimming off fat do not cut toward yourself, but in the opposite direction.

When finished, you'll have an oblong shell roast, trimmed to the meat along one edge, the rest with ¼ inch of fat remaining on it. Score the fat in any manner you choose before roasting.

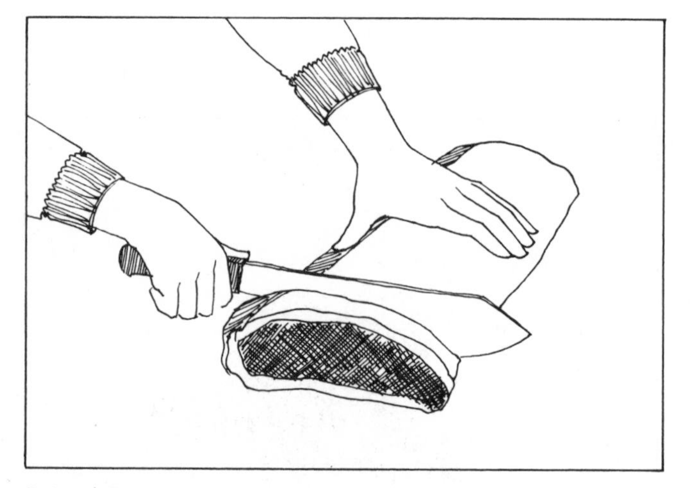

B-84. BONELESS SHELL STEAKS

BONELESS SHELL STEAKS

From the boned shell roast, cut down to preferred thickness for boneless shell steaks (B-84).

TAIL OF SHELL

Trim and use for ground meat.

FLANK

FLANK STEAK

Of the flank piece, 60 percent is waste. For flank steak, we recommend buying the trimmed flank from your supplier and using as is, or trimming as necessary.

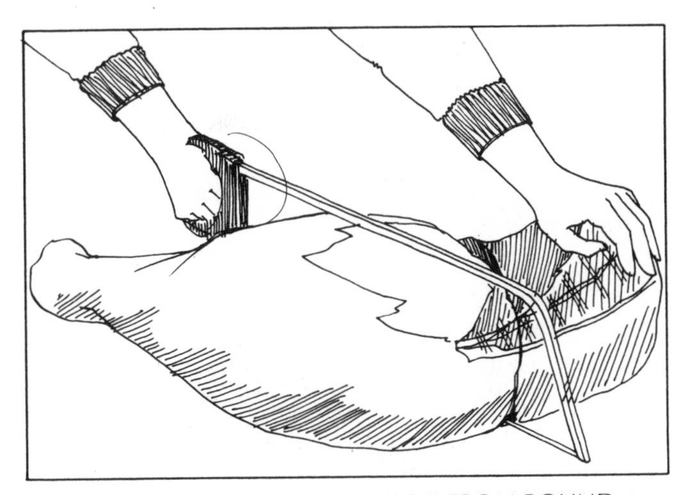

B-85. SAWING SIRLOIN-HIP SECTION FROM ROUND

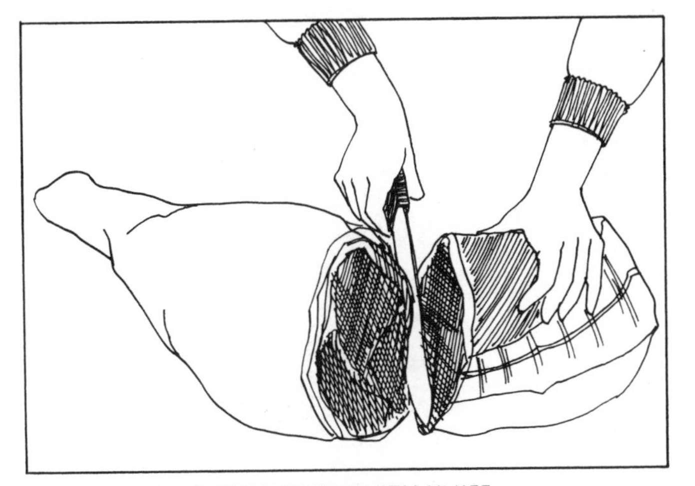

B-86. FINISHING SEPARATION WITH KNIFE

HIP (ROUND AND SIRLOIN)

The hip round is the entire piece. You can buy it in its entirety or already separated into the hip (sirloin section) and the round.

If you have bought the entire piece, you must first separate the hip (sirloin section) from the rest. You will see the long, cylindrical and curving hipbone (B–85). Measure approximately 1 inch from the bone and saw down entirely through the bone, finishing with the knife. Now the two sections are separated (B–86).

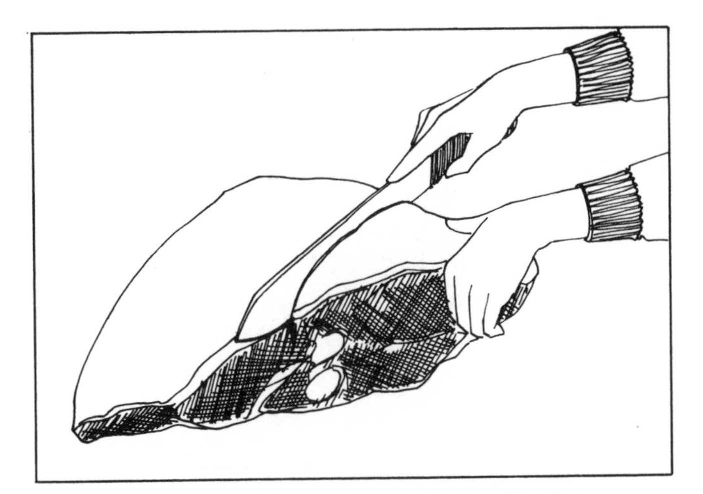

B-87. REMOVING TRIANGLE PIECE FROM SIRLOIN

SIRLOIN AND HIP SECTION

From this section you will get:

Sirloin Tip Roast (Silver Tip Roast)
Sirloin Steaks
Meat for Kabobs

First, cut off the tip section. Lay the entire piece down fat side up. On one end you will see a shiny round bone. Measuring on that end, place your knife into the meat directly over the round bone and cut down through the very center of the meat, using the 12-inch knife.

When you reach the top of the round bone, cut the meat in a triangle (B–87), which is the natural shape itself. There will be no bone in the section you cut away. You will have a triangular section of beef resembling a small Rock of Gibraltar (B–88).

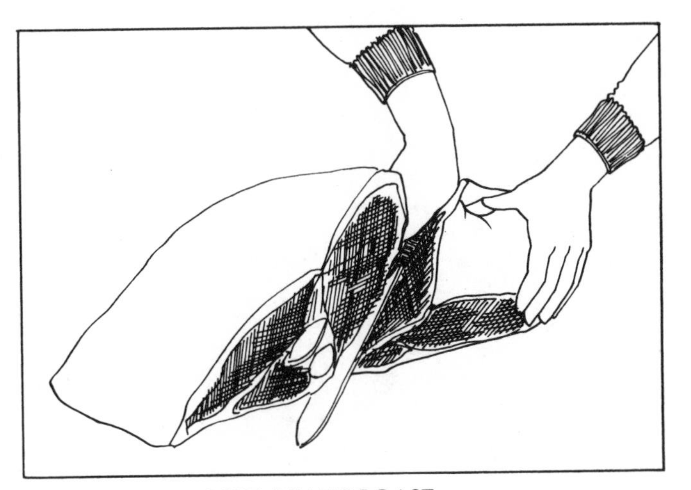

B-88. REMOVING SIRLOIN TIP ROAST

SIRLOIN TIP ROAST

This triangle is your sirloin tip roast. Trim the excess fat and roast.

What is left is the main sirloin or hip section. From this, four kinds of sirloin can be cut. Many people are unaware that there are three varieties of bone-in sirloin and one boneless. These are: *hip* or *pin-bone, flat bone, round bone,* and *boneless.*

We feel the two best are the flat and the round-bone pieces. The boneless section at the end is somewhat more sinewy.

When cutting, use the knife first, then the saw to go through the bone, finishing with the knife (B–89, B–90, B–91, and B–92).

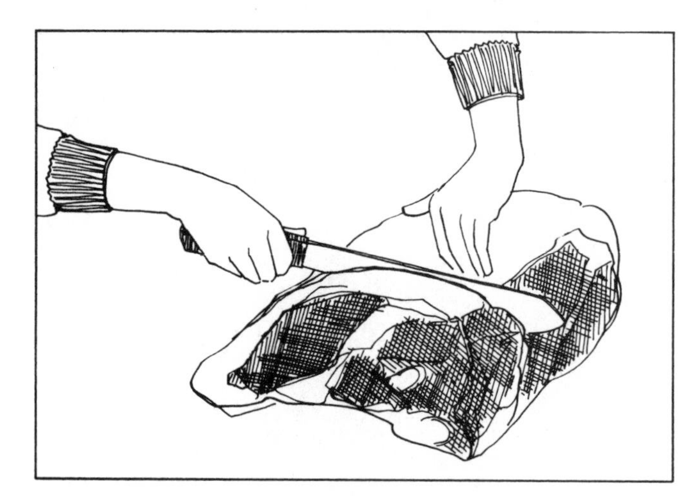

B–89. CUTTING PIN-BONE STEAKS

SIRLOIN STEAKS

Start at the end where you see the fillet section. That is the hip or pin-bone end. From the average sirloin section you should be able to cut yourself 1 hip or pin-bone steak, 2 flat-bone steaks, and 2 round-bone steaks, and have the boneless sirloin end left. Cut according to your preference for thickness and trim and score the fat before broiling.

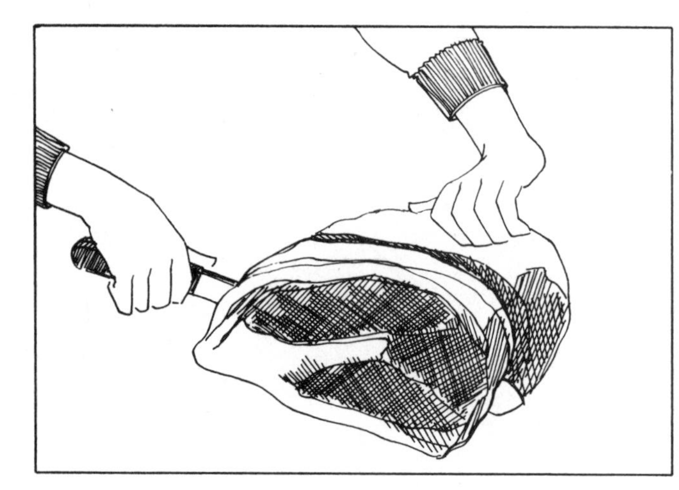

B–90. ROUND-BONE STEAKS

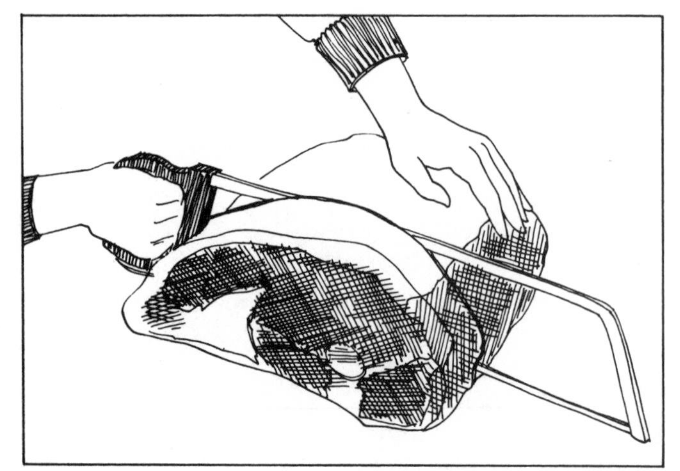

B–91. FLAT-BONE STEAKS

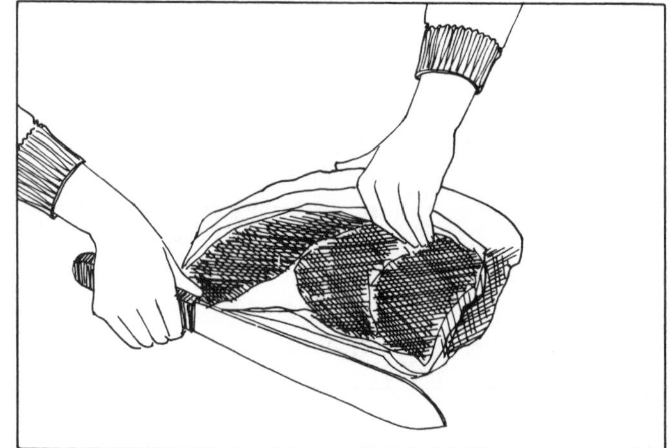

B–92. BONELESS STEAKS

THE ROUND

This is the other part of the entire hip round piece that you have separated. (Remember, this section can be purchased already separated from the hip sirloin section.) From this section you will get the meat for:

Top Sirloin
Top Sirloin Steaks
Pot Roast
Eye Round Roast
Elegant Pot Roast
Top Round
Top Round Roast Beef
London Broil
Stroganoff
Minute Steaks
Round Steak
Steak Tartare

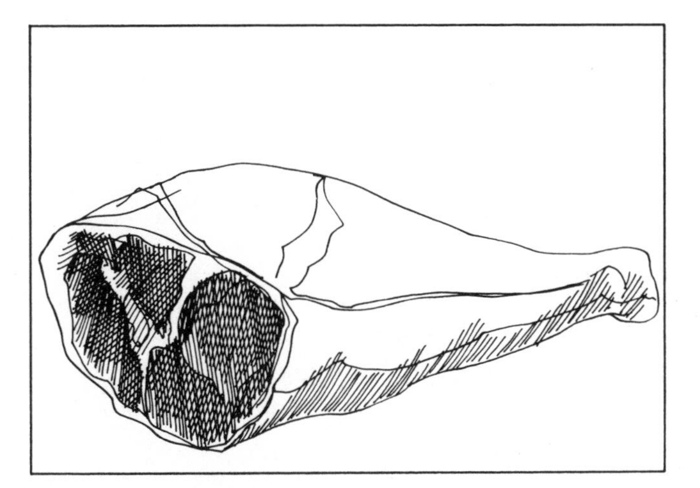

B-93. ROUND SECTION

Put the round section down with the fat side up, the meat facing you. You will see the round piece of ball bone (B-93). From the ball bone, using the tip of the knife, draw a line directly up to the top fat of the piece, and then measure approximately 1 inch to the left of this line. Make a cut here and you will see a small separation in the meat (B-94). Continue to make small, angled cuts into this separation. When you push your fingers into the opening (B-95), you will see that it follows down from the top of the meat at an angle. Continue to cut into this separation, widening it.

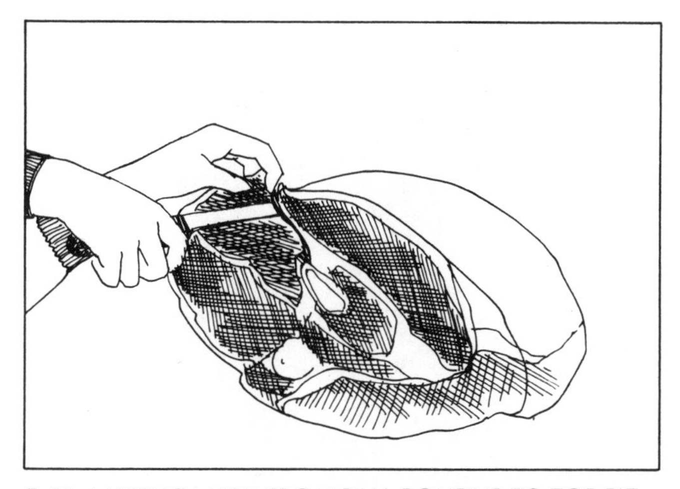

B-94. MAKING MARK FROM BALL BONE UP TO TOP FAT

The separation will keep opening at a widening angle, and running down the side of the entire piece. It will go right to the side and then all the way along the side to the knee-bone joint.

Continue to use your fingers to widen the separation. This is the preliminary work needed to remove the top sirloin section. Insert the knife as far as it will go—all the way to the hilt if need be. You now have a deep cut into this separation, ending at the knee bone. When you see this piece of smooth, shiny bone, make a cut with the knife at right angles to the other cut, into the side (B-96), right up against the knee bone. This right-angle cut is to act as a guide.

Now flip the entire round on the other side. The side you were cutting is now away from you (the knee joint). Find the mark you made on that side—that right-angle cut you just finished making. From this

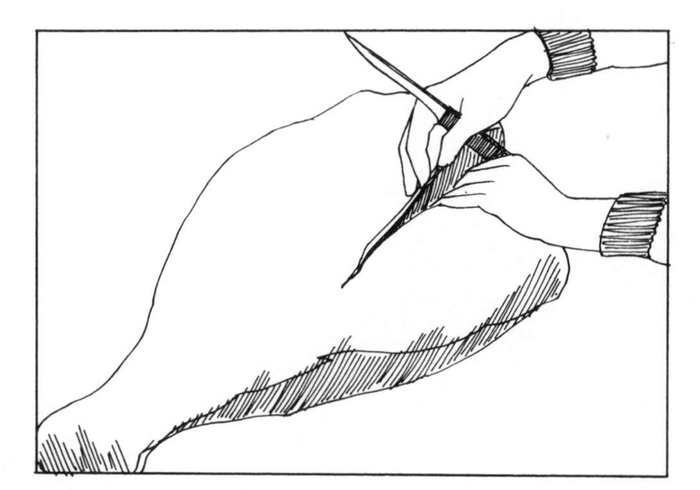

B-95. USING FINGERS TO SEPARATE MEAT

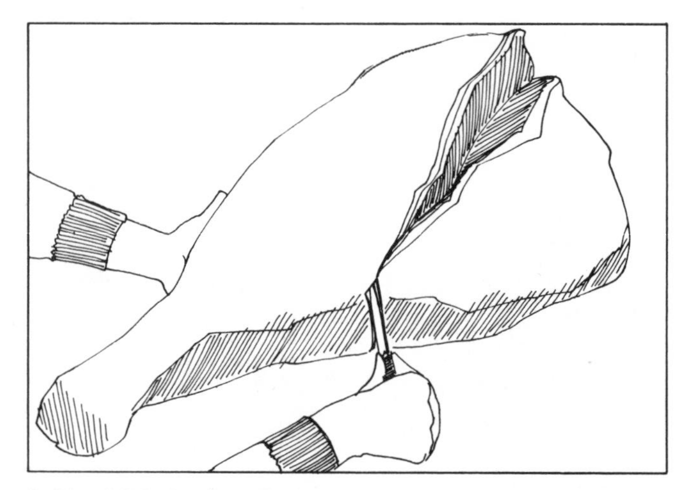

B-96. THE RIGHT-ANGLE CUT AGAINST KNEE BONE

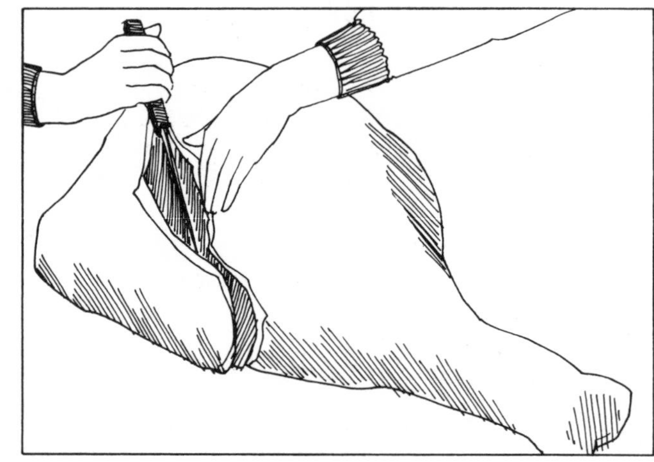

B-97. DIAGONAL CUT ALONG TOP OF LEG BONE

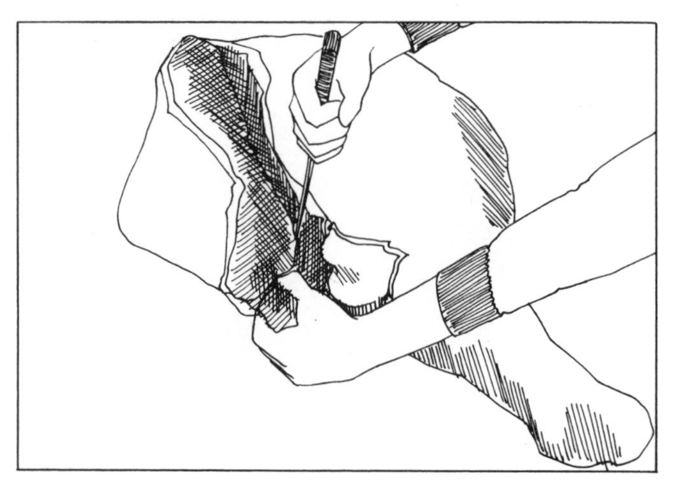

B-98. CUTTING WITH TIP OF BLADE ALONG TOP OF BONE

cut, you will cut at a diagonal angle to the top of the piece, precisely over the round ball-joint bone (shown in B–94) inside the meat (B–97).

The next step is to cut deep into the angle line you have just made. Cut slowly because your knife will have to go in deeply. Use the tip of the knife to feel the bone inside. You want to cut along the top of this bone, so keep touching it with the tip of the blade (B–98). The tip should actually always be moving along the top of the bone as you make this angled cut. When you finish the cut, you will come out at the face of the round ball bone in the meat (B–99).

This cut has now been deepened and opened. Next, go back in again, but this time cut not along the top of the bone but *around* the bone, pulling the meat away with your other hand as you do. Keep cutting along the side of the bone and take off the section of meat until you have removed it entirely (B–100). This top sirloin will be a triangular piece of meat.

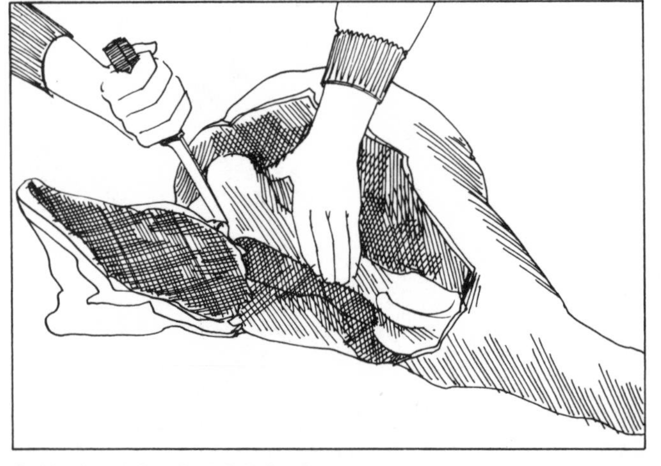

B-99. EXPOSING LEG BONE

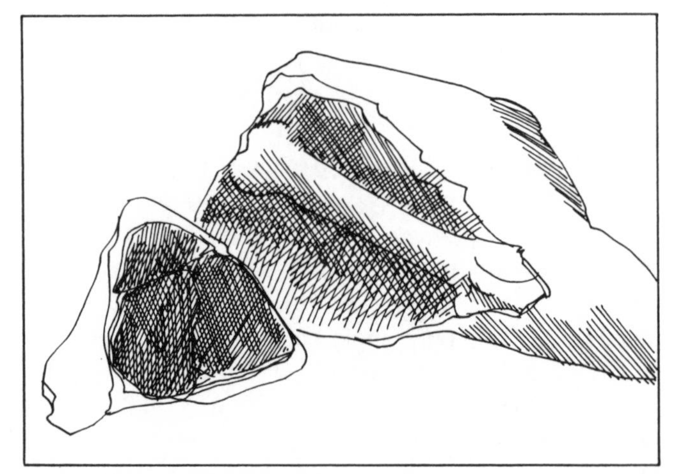

B-100. REMOVING TOP SIRLOIN SECTION

TOP SIRLOIN ROAST

Trim off all fat and the small, triangle-shaped piece at the very top, an overhanging piece with a fatty side (keep for ground or stew meat). Trim away the top membranes still on the meat to complete the top sirloin roast.

TOP SIRLOIN STEAKS

Cut the top sirloin roast piece on the bias for 2 top sirloin steaks and use the balance for a smaller roast.

To get the most out of the rest of the round section, you must bone out the entire piece.

Turn the end of the leg toward you and you will see the Achilles heel tendon on the leg. Just above this tendon, where the meat of the round starts, make a cut. Cut past the meat into the membrane. Make the cut about 4 inches from the hipbone (B–101).

From this small 4-inch cut, mark a line at an upward angle to the shiny elbow bone. Making small cuts, travel up along this line toward the elbow bone (B–102), following the line of the leg bone to the elbow bone and working around the inner edge of the bone. In doing this, you will need to cut through more than one layer of tendon. Work slowly with small, scraping cuts right along the bone. Follow on up along the entire length of the bone, all the way to the smooth ball end, pulling the meat back as far as you can (B–103).

Now turn the entire section over. Return to the cut you made just over the Achilles heel tendon and expand the cut from this other side down to the shinbone (B–104).

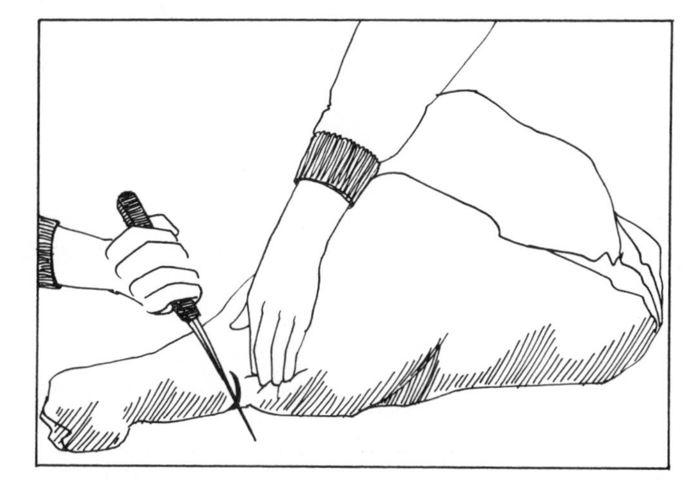

B–101. INCISION ABOVE ACHILLES HEEL

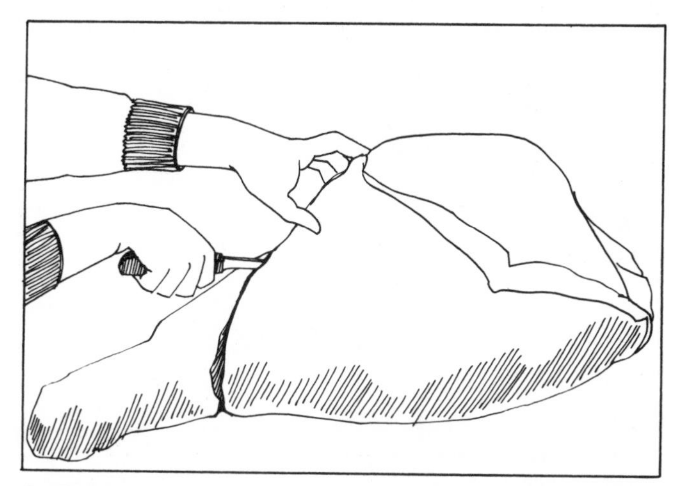

B–102. SMALL CUTS FOLLOWING LINE OF LEG BONE TO ELBOW BONE

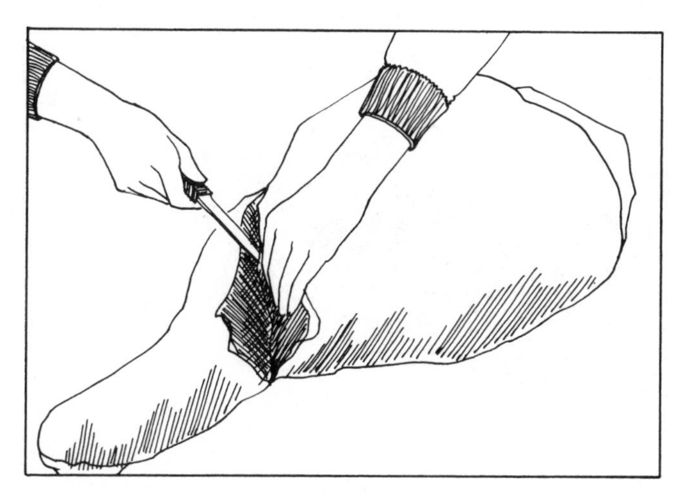

B–103. WIDENING INCISION USING HAND

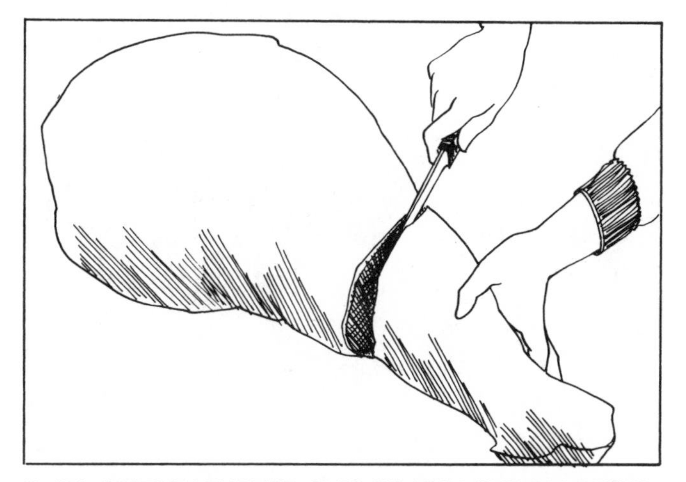

B–104. SECTION TURNED OVER TO JOIN CUT ON OTHER SIDE TO SHINBONE

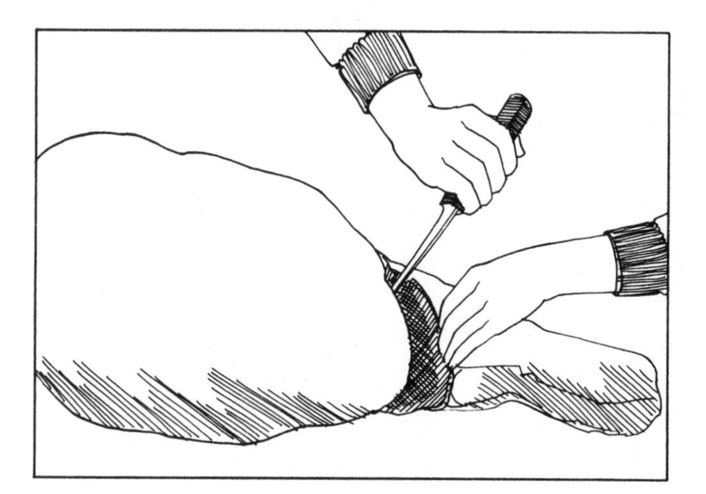

B–105. SHIN BEEF OVER BONE

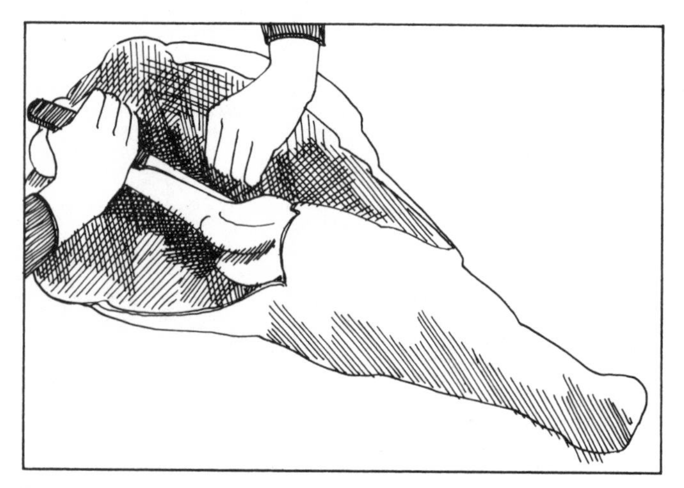

B–106. USING KNIFE AND HAND TO FOLLOW LINE OF BONE

Now make another angled cut mark ending up at the elbow bone as you did with the other side and cut in again along this angled mark, following the contours of the leg. As you cut the tissue away you'll see the shin beef over the bone on this side (B–105). Keep following the upward contour of the leg and then along the entire length of the leg bone (B–106). *Note:* Use the tip of your knife to make small cuts or you will cut into the meat too deeply.

The entire leg bone is now almost free. You must cut around the ball and socket of the upper end and free this completely. Using the tip of the knife, keep working around the joint, as close as you can, simply following the contours of the bone (B–107 and B–108). Boning the piece in this manner requires the least amount of lifting and turning.

Now turn the piece with the fat side up. The meat facing you is bottom round. This has to be sectioned out of the rest of the piece. Start at the top of the corner nearest you and make a small cut while pulling on the thick top flap of meat with your other hand (B–109). Cut into the seam that appears as you pull back the top of the thick covering. Keep cutting and pulling. The thick top piece will be coming up in your other hand. Keep following the natural division, or seam, as it appears (B–110).

As you cut deeper, keep holding this large piece with your other hand, as it will become floppy. Finally, when you've cut through the seam and the piece is about ready to come off, you will have a deep V cut. Make the final cut at this end and take off the bottom round piece (B–111).

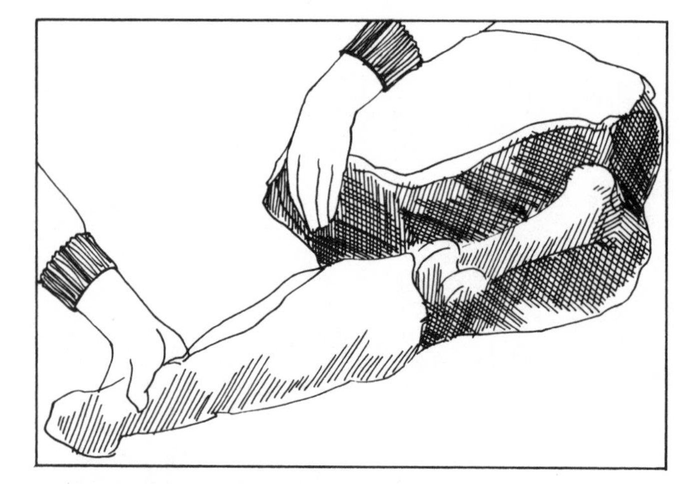

B–107. PULLING BONE AND SHANK AWAY FROM MEAT

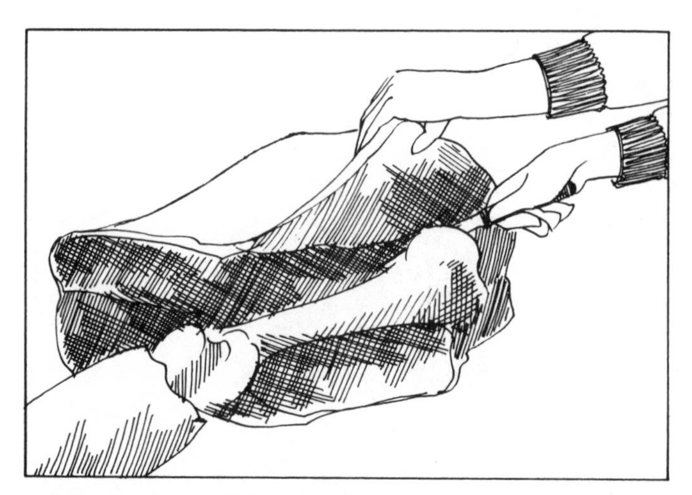

B–108. CUTTING OUT BONE AT JOINT

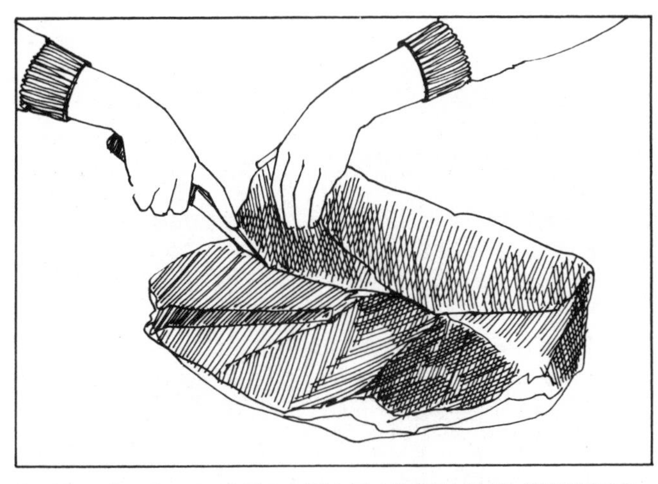

B-109. USING HAND TO PULL FLAP OF MEAT AWAY FOR BOTTOM ROUND

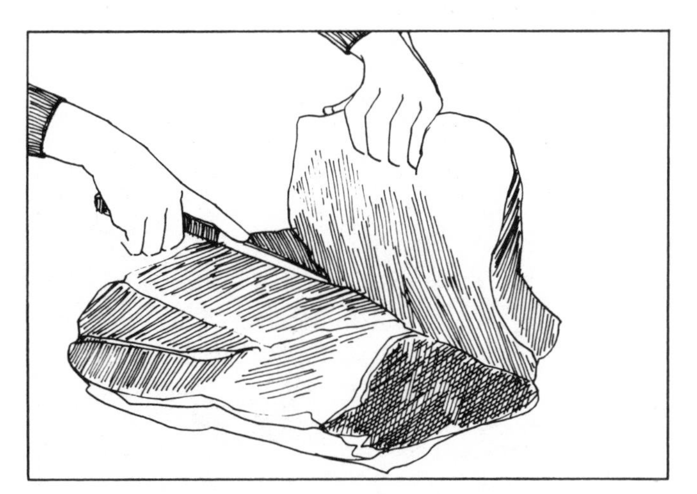

B-110. FOLLOWING NATURAL SEAM

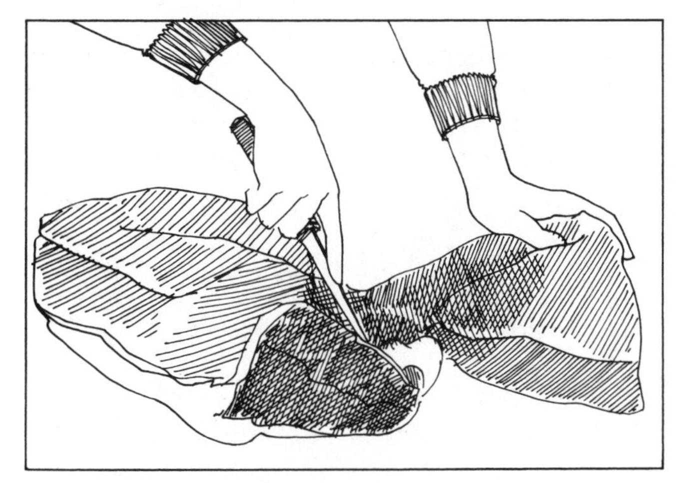

B-111. REMOVING ENTIRE BOTTOM ROUND

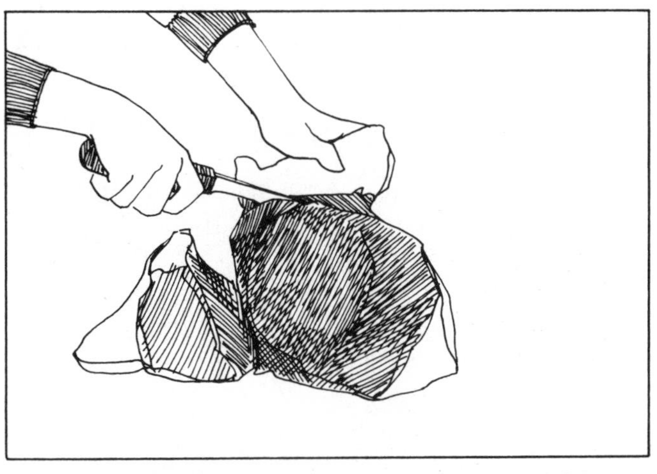

B-112. TRIMMING FAT. PIECE MAY BE LEFT WHOLE OR CUT INTO HALVES OR THIRDS

POT ROAST

Trim off the thick membranes along the side of the piece, along with any fat. Use the piece in its entirety for pot roast, or cut into halves or thirds (B–112).

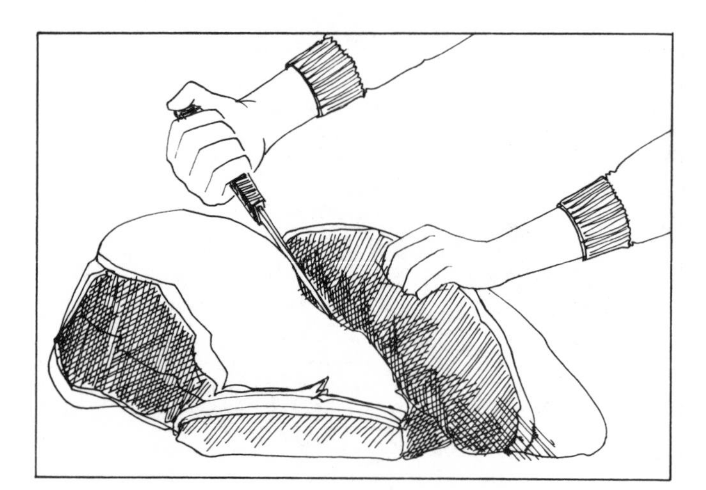

B-113. CUTTING INTO "VALLEY" FOR EYE ROUND

EYE ROUND

With the fat side of the rest of the piece facing you, take out the eye round as follows. Note the deep and obvious valley in the piece of meat. Cutting into this valley, or seam (B–113), you will soon be able to begin pulling the two pieces of meat apart by hand. Then, keep making small cuts into the natural seam to separate the eye round (B–114). Pull the eye round back almost over onto itself and continue cutting in and under, until you have taken off the entire piece (B–115). This separated piece of eye round is thick and cylindrical in shape.

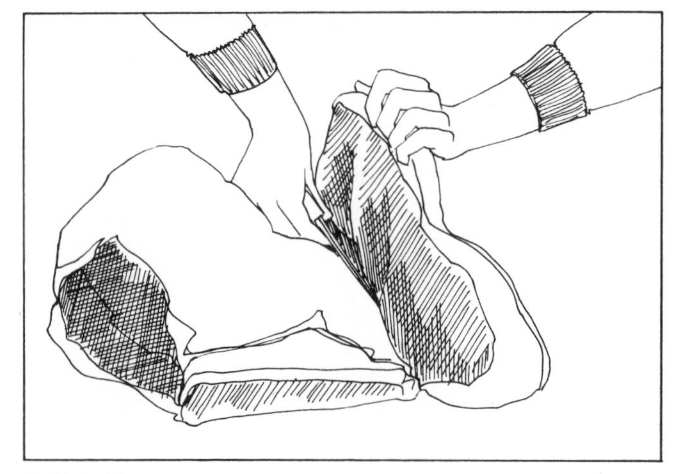

B-114. SEPARATING SEAM TO PULL OUT EYE ROUND

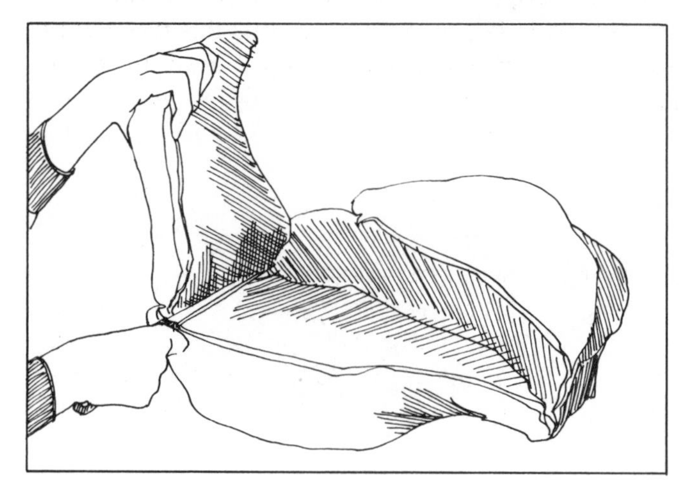

B-115. REMOVING EYE ROUND

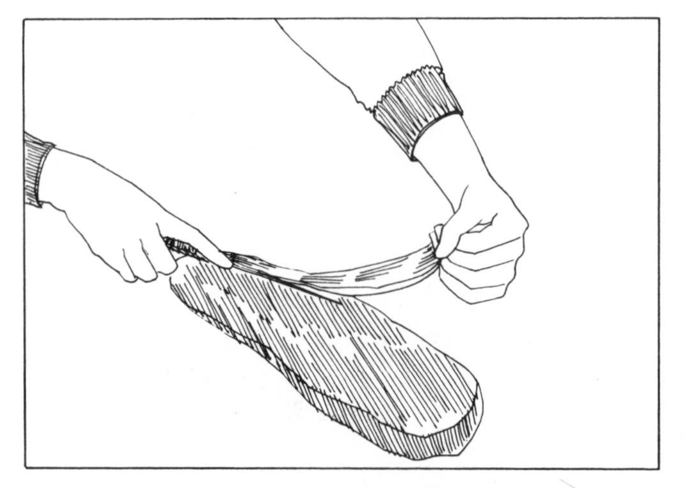

B-116. TRIMMING SINEWS FOR EYE ROUND ROAST

EYE ROUND ROAST

Remove all the sinews and fat still covering the piece (B–116). However, as eye round has little graining inside, it will be necessary to put some of the fat back on top in loose layers before tying and roasting.

ELEGANT POT ROAST

If the eye round is cut into thirds, one of these pieces makes an elegant pot roast because it is almost without fat. This cut is perfect for pot roast for those who must watch their cholesterol intake.

TOP ROUND

The top round comes from the remaining piece. Turn the entire piece over with the fat side up. You will see a narrow, slightly curving bone (the pelvis bone), which must be removed (B–117). Cut around this bone on all sides and pull it out in its entirety (B–118).

Lay the remaining piece of meat fat side down. You'll see a thick section on the top with one end shaped a little like a ship's prow. Pull this piece back with your hand and begin cutting it away from the larger, main piece of meat (B–119). Do this by cutting through the thin layer of connecting tissue between this piece and the main section.

After you have cut off the top piece, trim away all fat and membranes. This is shin meat; the shin is discussed further at the end of the chapter.

The large, thick, squarish piece left is the top round. Trim away any excess bits and ends of fat hanging from it. Trim the thick, *outer* fat from this sizable piece of meat, leaving on a thin layer (B–120).

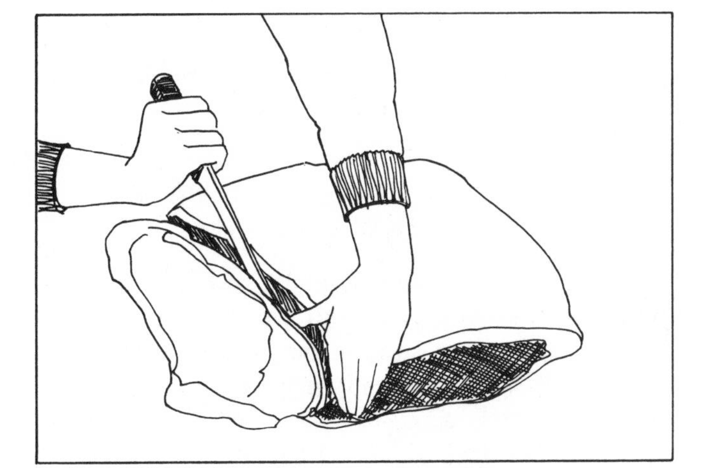

B–117. INITIAL CUT TO REMOVE PELVIS BONE

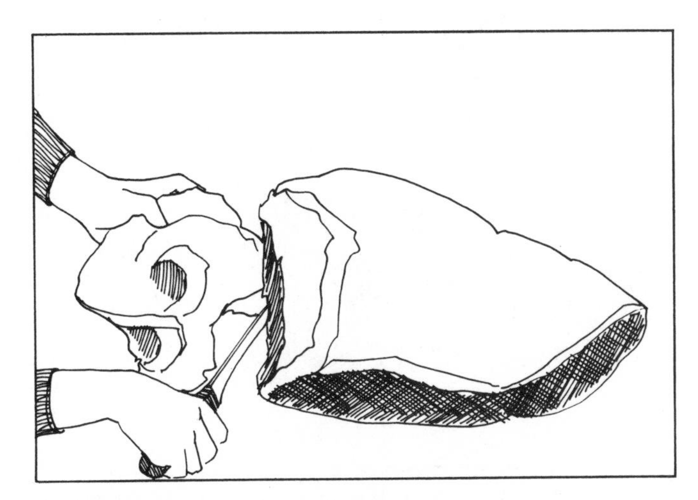

B–118. REMOVING PELVIS BONE

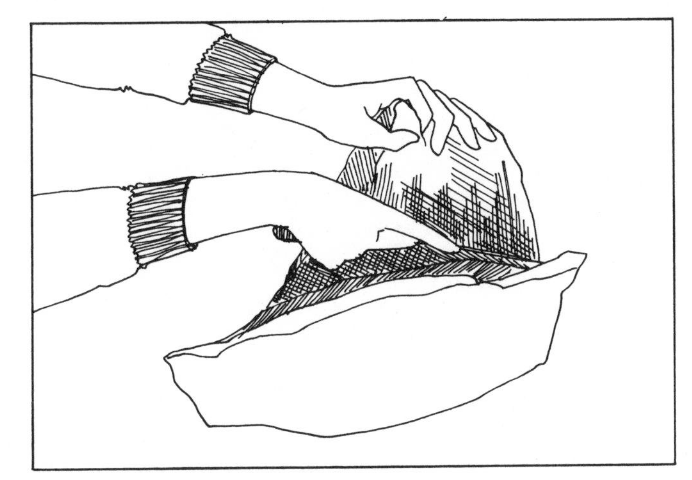

B–119. USING HAND TO AID IN SEPARATING MEAT

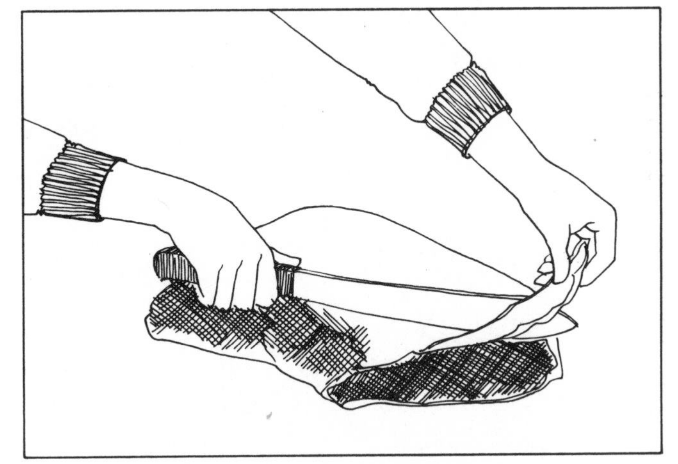

B–120. TRIMMING FAT FROM TOP ROUND

ROAST BEEF

The whole top round makes a fine roast beef for 18 to 20 people. Cut in half, it makes a roast for 10 people.

LONDON BROIL

The top round, though squarish, comes to a point in the rear. The broad end can be used for London broil. Go back about 6 inches and retain the first-cut London broil. From this end, cut on the bias (B–121) for approximately six 1-inch London broil strips or twelve ½-inch strips. Use the rest of the top round for pot roast, roast beef, or a lean stew.

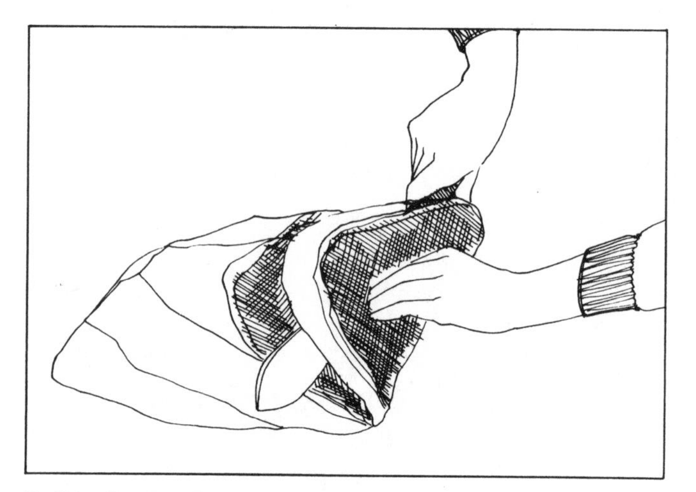

B–121. CUTTING LONDON BROIL FROM TOP ROUND

STROGANOFF

Using the first cut of the top round, cut across the grain and on the bias for beef Stroganoff strips. Cut the strips ½ inch thick.

MINUTE STEAKS

You obtain these by cutting very thin slices from the first cut of the top round piece.

ROUND STEAK

Round steak is cut from the top round piece, usually 1 to 3 inches thick. Simply cut down to the desired thickness on your top round piece. We suggest round steak be marinated before broiling.

STEAK TARTARE

For steak tartare use the lean section near the rear, past the first cut. Make certain you have trimmed and removed all fat. Grind or mince fine for steak tartare.

SHIN

The shin section will still be left. Just follow the line of the bone and remove all the meat you can comfortably take off. When freeing the meat from the bone, work the knife close to the bone. After the meat is off the bone, trim away all fat, sinews, etc., which will be clearly visible. Use the meat for grinding and the bones for stock.

9

Variety Meats

For these pieces you will have to do little cutting. For the most part, brains, tripe, liver, kidneys, etc., are bought from your wholesaler or supplier already separated from the animal. However, on occasion you may find the kidneys or liver still in with the part of the carcass you have purchased. In most instances, it will probably be the kidneys so we'll start with those.

Beef and veal kidneys are given a slightly lumpy appearance by the lobes. Pork and lamb kidneys are smooth. Veal kidneys will be paler in color than the others. Check the odor of the kidneys, to make certain they are fresh. If they are still enclosed in the section of meat you've purchased, cut them out carefully, trimming away all excess fat around them. Sometimes this is a considerable amount. Kidneys are covered with a membrane or sheathing that must be removed when you take them from the carcass. Use a very sharp 6- or 8-inch knife, working in with the very tip first, to lift up a corner of the membrane.

When the outer membrane has been removed, trim the kidneys. With the large beef and veal kidneys, you will see the fatty line easily. Cut into its center and down into the middle of the kidney, where you'll see a hard center. Core this out as though you were coring an apple.

All kidneys should be parboiled before they are cooked in any other way. Kidneys are highly perishable, as are most organ meats, and should be used within 2 days of purchase or removal from the carcass. They should be refrigerated at once. Fresh kidneys may be frozen if wrapped carefully. In a self-contained freezing unit, they can be stored for up to 3 months.

LIVER

When you bring liver home it should be trimmed at once. You will see a thin, outer skin, cartilege, and small veins. Remove all these, using the point of the knife to make your initial start. Do not cut in, but keep the knife flat along the surface of the liver, using small, gliding strokes, lifting and cutting, until you have removed all the outer membranes, etc. Rinse the liver in room-temperature water and refrigerate at once. Use within 2 days. We do not favor freezing liver.

CERVELLE (BRAINS)

These have always been much more popular in Europe than in America; many restaurants in this country still use the French word for brains in describing dishes made with them. Of all organ meats, these are the most delicate in taste and texture. They are also among the most perishable and should not be kept in a refrigerator for more than 2 days. They can be frozen, if carefully wrapped, in a self-contained freezer unit for up to 3 months.

Beef and veal brains are the most popular. Your supplier has probably already removed them from the carcass and taken off the covering membrane. Small bits of membrane may remain, however, and should be removed. It is easiest to do this after blanching or soaking. Cover the *cervelle* with water to which you have added 1 to 1½ teaspoons salt and the juice of a quarter of a lemon. Soak the brains in this mixture in the refrigerator for 2 hours, or simmer gently in the mixture for 20 to 30 minutes. If simmering, drain when finished and cover again with cold water. Any bits and pieces of membrane remaining can be easily removed with the tip of the knife.

HEART

You will probably also purchase hearts from your supplier already removed from the carcass rather than doing this yourself. Heart meat is highly nutritious but fairly tough, and moist-heat, long-cooking methods such as pot roasting or braising must be employed. Beef heart is, naturally, the largest, weighing up to 5 pounds. Veal heart is next in size, weighing 1 to 1½ pounds, with pork and lamb hearts the smallest, weighing in the 5-to-9-ounce range.

Heart can be sliced or diced for use in stews. Another use for heart is to grind it with chuck or round to make an especially nutritious hamburger.

SWEETBREADS

Also very popular in Europe and in the American Southland, sweetbreads are perhaps the most versatile of all organ meats. They are the thymus glands of young animals. Young beef sweetbreads are usually a slightly grayish color, as are veal sweetbreads. Bulk suppliers sell them in what is called a "cluster," consisting of 2 lobes, a heart sweetbread and a throat sweetbread. This cluster or "pair," as it is sometimes called, of good, fresh sweetbreads should weigh from ¾ to 1½ pounds. Sweetbreads weighing from ¼ to ¾ of a pound are pieces and far less desirable than a whole cluster.

Sweetbreads are as perishable as brains, with a maximum refrigerator-storage time of 2 days. Properly wrapped, they can be stored for up to 6 months in a self-contained freezer unit.

Examine the sweetbreads when you get them home. If your supplier has trimmed them, there is little for you to do. If he has not, and this may often be so with a bulk supplier, some blood vessels or connective tissue may still remain. Soak them first in cold water, changing the water if necessary, until all the blood is gone from the sweetbreads. Then, using the point of a knife, carefully remove blood vessels, connective tissue, and outer skin. You will see a delicate inner membrane underneath. Leave this on as it keeps the sweetbreads together. They are now ready to cook according to your recipe.

TRIPE

Once more, Europeans are more familiar with tripe than we are. Tripe is the stomachs of cattle, which, unlike humans, have more than one stomach. There are two kinds of tripe, the plain or smooth kind and the "honeycomb" tripe.

Tripe requires a great deal of careful preparation by professionals at the meat-packing plants, including a thorough cleaning and scalding of the paunches. Tripe is, therefore, partially cooked when you buy it, though it is called "uncooked." In any case, it requires a great deal more cooking before it is edible. Tripe is not something we suggest you try to prepare from scratch but purchase ready for use from your supplier.

TONGUE

If you decide to purchase fresh beef or veal tongue from your bulk supplier, use the tongue within 3 or 4 days. Fresh tongues may be stored frozen for 6 months. They should be frozen *before* cooking, and, of course, carefully wrapped.

A fresh tongue, as it comes from the market, still has the skin on it, a little fat at the back end, and, at the same end, some little bones attached to the root. It is easier to remove these items after you parboil the tongue. Add 1 teaspoon salt to each quart of water, immerse the tongue, and parboil it for approximately 30 minutes, depending on the size of the tongue. When you've finished parboiling, drain off the water, let the tongue cool, and then begin to remove the skin, whatever fat there is, and the little bones. We should note here that some people prefer not to remove the little root bones until the tongue is completely done because when they pull out easily, this is a sign that the tongue is thoroughly cooked.

Begin removing the skin on the *underside* of the tongue. Make a little slit with the knife, cutting carefully with little slits, holding the tip of your knife fairly flat so as not to cut into the tongue. Peel back the skin as you go toward the tip of the tongue. Reaching the tip, you may have to use your knife to cut off the part of the skin you've already loosened. If not, just continue peeling back over the top, removing the entire skin as you would a long glove.

A final note on your fresh tongue purchase: Beef tongues usually weigh in at 2 to 5 pounds, averaging 3½ pounds; veal tongues weigh in at from ½ to 2 pounds. *Note:* The parboiling described above does not cook your tongue, but only assists you in removing the skin, fat, etc. Cooking, simmering, or poaching a beef tongue takes from 3 to 3½ hours; a veal tongue from 1¾ to 2¼ hours.

10 Freezer Facts

Now that you know about the cutting, let's go into the keeping. Use of the home freezer is an important part of the techniques you have mastered in cutting your own meat. Freezing is the only method by which meat can be preserved in a condition close to its normal or fresh state. If you freeze meat properly, your larder will be ready and waiting to release all its locked-in flavor when you want it. If you freeze it improperly, all your good work will be for nothing, because incorrectly packaged and frozen meat will dry out and lose flavor and nutrition.

You should know what takes place when you freeze meat. Meat begins to freeze at about 29°F. As the temperature drops, the growth of tiny microorganisms comes to a halt, and the hydrolytic enzyme action on protein and fat becomes minimal. Because of the greater exposure to air, the surface of a piece of meat always oxidizes more rapidly than the inside, and different animal fats oxidize at different rates. Beef and lamb, for example, do so more slowly than pork.

Most freezer storage is at temperatures of 0°F, or less. Certainly your freezer should never be above 10°F. Freezing involves a drying process, which changes the composition of the meat. The meat juices are not re-formed exactly as they were when the meat thaws. But if the freezing is done quickly and properly, the changes in the composition will be so slight as to be unnoticeable to the palate. When meat is frozen improperly (too slowly), and stored incorrectly, the changes in composition become greater, with a resultant loss of flavor and quality.

Freezer burn occurs because the air in freezers is dry (freezer coils extract most of the moisture from the air). The dry, freezing air circulates from the coils to the rest of the unit, around the frozen food, absorbing all the moisture there is. Incorrectly wrapped meat dries out because the cold air removes the protective ice crystals from the meat that first form as the meat freezes. The result is freezer burn: a dry, pitted, discolored surface; a hard, dry piece of meat with unappetizing flavor. It may be preserved, but its composition has changed and it will never recapture the "fresh-meat" flavor which is the mark of thawed meat when properly frozen.

Wrapping your meat is, obviously, of primary importance. Try not to wrap pieces that are too large and unwieldy. Wrap with a high grade of freezer paper—there are numerous brands on the market. Double-wrap each piece. Aluminum foil may be used, but we prefer the paper specially made for freezer wrapping. Aluminum foil has a tendency to become brittle after prolonged exposure to low temperatures and it can develop tears or holes, which will, of course, undo all your careful wrapping. Ordinary wax paper is not satisfactory for freezer wrapping.

Examine the piece of meat you are wrapping. If there are sharp corners, dull them or cut them off. Wrap the initial paper tightly, smoothing it down on all sides and then folding the sides and edges over to press out the air. Then cover with a second wrap. After you've wrapped your packages properly, tie, heat-seal or seal with freezer tape. A note of warning on heat-sealing. This can be a tricky process and, unless it is done carefully, will not give you the seal you need. Proper sealing with freezer tape can be less uncertain in the long run.

Labeling is the next important step. Do not fool yourself into thinking that you will be able to recognize all your frozen pieces by their shape, especially after weeks and months. Label the pieces you store in your freezer—and *date* them. Do not trust to your memory. Be sure to use the tapes that are specially made for freezer labeling. Ordinary tape comes off as it dries out in the very low temperatures of your freezer.

How long can meat be frozen? The Department of Agriculture recommends these periods:

Beef—12 months
Veal—12 months
Lamb—12 months
Fresh pork—8 months
Ground beef—4 months
Ground lamb—4 months
Pork sausage—3 months

We find this too general, too unspecific, because it implies freezing under the most professional of conditions for the freezing unit itself as well as the wrapping and storing. We suggest the following guide as being more realistic for the home-freezing schedule. We prefer to err on the safe side:

	UNCOOKED	COOKED
BEEF		
Large pieces	6 months	3 months
Small pieces	4 months	2 months
Marrow bones	6 months	Do not freeze
Corned Beef	1 month	2 months
Ground beef	3 months	Do not freeze
Beef fat	6 months	Do not freeze
VEAL		
Large pieces	4 months	2 months
Small pieces (chops, etc.)	1 month	3 weeks
Ground veal	1 month	Do not freeze
LAMB		
Large pieces	6 months	3 months
Shanks	4 months	3 months
Chops	4 months	1 month
Ground lamb	3 months	Do not freeze
FRESH PORK		
Large pieces	4 months	3 months
Chops, spareribs	3 months	Do not freeze
Ground pork	3 months	Do not freeze
CURED PORK		
Ham (whole or half)	3 months	Do not freeze
Ham steaks	1 month	Do not freeze
Bacon (sliced in sealed package)	1 month	1 month
Bacon (slab)	1 month	1 month

CHICKEN		
Whole	6 months	1 month
Cut up	3 months	1 month
TURKEY		
Whole	6 months	3 months
Cut up	3 months	1 month
DUCK		
Whole	6 months	1 month
GOOSE		
Whole	6 months	3 months
QUAIL		
Whole	2 months	1 month
WOODCOCK		
Whole	2 months	1 month
ROCK CORNISH HENS		
Whole	2 months	1 month
ALL POULTRY GIZZARDS	3 months	2 months
POULTRY LIVERS	2 months	Do not freeze
VARIETY MEATS		
Kidneys	6 months	1 month
Liver	3 months	Do not freeze
Hearts	6 months	3 months
Sweetbreads	4 months	Do not freeze
Brains	3 months	Do not freeze
Tripe	3 months	1 month
Tongue (fresh)	6 months	4 months

Of course, the last part of the freezing process is properly unfreezing (defrosting) your meat. Use defrosted meat as quickly as possible after the thawing process is complete so bacteria are not given a chance to grow. The refrigerator is the best place to defrost it. In a way, thawing is the method through which your meat re-forms itself as close as possible to its original, fresh state. Thawing allows the meat to reabsorb the frozen moisture and particles of juice.

Sometimes, the piece of meat is too large to defrost in your refrigerator. Meat can be allowed to thaw at room temperature, but close watch should be kept on the progress of the defrosting, so that when it is finished, the meat can be cooked at once. If time is a factor in thawing, defrosting can be done in warm water, preferably running, with the meat in the original package. As a general rule, it will take 3 to 4 hours per pound to thaw meat inside the refrigerator, 2 to 2¼ hours per pound at room temperature, and 1 to 1½ hours per pound in warm water.

Properly frozen and wrapped, and properly thawed, meat will retain flavor and nutritive value to an amazing degree.

INDEX